VOLUME ONE HUNDRED AND FIFTY FOUR

PROGRESS IN MOLECULAR BIOLOGY AND TRANSLATIONAL SCIENCE

Sirtuins in Health and Disease

VOLUME ONE HUNDRED AND FIFTY FOUR

PROGRESS IN MOLECULAR BIOLOGY AND TRANSLATIONAL SCIENCE

Sirtuins in Health and Disease

Edited by

WEIPING ZHENG

School of Pharmacy, Jiangsu University, Zhenjiang, People's Republic of China

ELSEVIER

ACADEMIC PRESS

An imprint of Elsevier

Academic Press is an imprint of Elsevier
125 London Wall, London EC2Y 5AS, United Kingdom
525 B Street, Suite 1800, San Diego, CA 92101-4495, United States
50 Hampshire Street, 5th Floor, Cambridge, MA 02139, United States
The Boulevard, Langford Lane, Kidlington, Oxford OX5 1GB, United Kingdom

First edition 2018

ISBN: 978-0-12-812261-7
ISSN: 1877-1173

For information on all Academic Press publications
visit our website at https://www.elsevier.com/books-and-journals

Publisher: Zoe Kruze
Acquisition Editor: Ashlie M. Jackman
Editorial Project Manager: Joanna Collett
Production Project Manager: James Selvam
Cover Designer: Mark Rogers

Typeset by Thomson Digital

CONTENTS

CONTRIBUTORS

Iacopo Galleano
Center for Biopharmaceuticals, University of Copenhagen, Copenhagen, Denmark

Xiao Hu
School of Pharmacy, Jiangsu University, Zhenjiang, People's Republic of China

Shengchao Li
School of Pharmacy, Jiangsu University, Zhenjiang, People's Republic of China

Andreas S. Madsen
Center for Biopharmaceuticals, University of Copenhagen, Copenhagen, Denmark

Sin Hui Neo
Yong Loo Lin School of Medicine, National University of Singapore, Singapore

Christian A. Olsen
Center for Biopharmaceuticals, University of Copenhagen, Copenhagen, Denmark

Nima Rajabi
Center for Biopharmaceuticals, University of Copenhagen, Copenhagen, Denmark

Anthony A. Sauve
Weill Cornell Medical College, New York, NY, United States

Bor Luen Tang
Yong Loo Lin School of Medicine, National University of Singapore, Singapore;
NUS Graduate School for Integrative Sciences and Engineering, National University of Singapore, Singapore

Ning Zhang
Weill Cornell Medical College, New York, NY, United States

Weiping Zheng
School of Pharmacy, Jiangsu University, Zhenjiang, People's Republic of China

PREFACE

Sirtuins refer to a family of physiologically and pathologically important intracellular enzymes present in all life forms on this planet. These enzymes catalyze the unique β-nicotinamide adenine dinucleotide (β-NAD^+)-dependent protein N^ε-acyl-lysine deacylation and protein adenosine diphosphate (ADP)-ribosyl transfer, two group transfer reactions. Since the discovery of the first sirtuin family member [i.e., the yeast silent information regulator 2 (sir2) protein] in 2000, the sirtuin field has undergone a steadfast growth, especially during the past few years, as highlighted by (1) the expansion of the substrate repertoire from histone proteins to many nonhistone proteins residing in various subcellular compartments inside eukaryotic cells and (2) the expansion of the N^ε-acyl-lysine's acyl repertoire from acetyl to bulkier ones removable more proficiently by many sirtuins, just name a couple. As such, the sirtuin-catalyzed deacylation has been increasingly demonstrated to play an important regulatory role in many crucial cellular processes, such as transcription, metabolism, and DNA damage repair. The sirtuin-catalyzed deacylation has also been regarded as a contemporary therapeutic target for human diseases, such as cancer, metabolic and neurodegenerative diseases. The current volume entitled "Sirtuins in Health and Disease" introduces to readers the N^ε-acyl-lysine substrate specificity, the regulation by the cellular metabolism of the cosubstrate β-NAD^+, the catalytic mechanism details and their elucidation (from a chemical biology standpoint in particular), the development of the inhibitory compounds for the sirtuin-catalyzed deacylation (the primary cause responsible for the cellular functions of sirtuins). In addition, one chapter on sirtuins and Huntington's disease (one typical neurodegenerative disease) is also included to highlight the therapeutic role of the sirtuin-catalyzed deacylation. Moreover, a chapter on SIRT4 and SIRT7, the two least-understood mammalian sirtuins is also presented. It is our hope that this volume will be appealing to those working on the biology (including biochemistry), physiology, medicinal chemistry, chemical biology, pharmacology, and bioorganic chemistry of sirtuins and the unique reactions they catalyze. It is also our hope that this volume will also benefit the researchers outside of the sirtuin field in that the lessons learned in sirtuin studies might also be applicable to other fields of study.

WEIPING ZHENG
School of Pharmacy, Jiangsu University,
Zhenjiang, China

CHAPTER ONE

Chemical Probes in Sirtuin Research

Xiao Hu, Weiping Zheng[1]
School of Pharmacy, Jiangsu University, Zhenjiang, People's Republic of China
[1] Corresponding author. E-mail address: wzheng@ujs.edu.cn

Contents

Abstract

Sirtuins refer to a family of intracellular enzymes that are the yeast silent information regulator 2 (sir2) protein homologs found in organisms from all the three kingdoms of life. This family of enzymes primarily catalyze the protein N^{ε}-acyl-lysine deacylation reaction despite the report for a type of bacterial/fungal sirtuins to robustly catalyze a protein mono-ADP-ribosylation reaction, however, these two group transfer reactions employ the redox coenzyme β-nicotinamide adenine dinucleotide (β-NAD^+) as the obligatory cosubstrate. Since 2000, in addition to histone proteins, more and more nonhistone proteins have also been identified as native substrates for the sirtuin-catalyzed deacylation, consistent with the ever-increased demonstration that this enzymatic reaction plays an important regulatory role in a variety of cellular processes, such as gene transcription and metabolism. This latter role is also consistent with the absolute dependence on β-NAD^+ of the deacylation reaction catalyzed by sirtuin family members. The sirtuin-catalyzed deacylation has further been proposed as a contemporary therapeutic target for human diseases, such as cancer, neurodegenerative and metabolic diseases. In order to fully tap the therapeutic potential of the sirtuin-catalyzed deacylation, the past few years have witnessed a tremendous advancement in mechanistic elucidation, chemical modulator (inhibitor and activator)

Progress in Molecular Biology and Translational Science, Volume 154
ISSN 1877-1173
http://dx.doi.org/10.1016/bs.pmbts.2017.11.014

development, (chemical) biological and pharmacological exploration of the sirtuin-catalyzed deacylation reaction. During the journey of this knowledge advancement, the use of carefully designed chemical probes has played an elegant role. This chapter will delineate the development and application of these chemical probes in sirtuin research.

1. INTRODUCTION

The word "sirtuin" is the name of a family of intracellular enzymes that are found in the organisms from all the three kingdoms of life (i.e., bacteria, archaea, and eukarya) and are the homologs of the yeast silent information regulator 2 (sir2) protein, the founding member of the sirtuin family.[1,2] As such, the other sirtuin family members were initially thought to also possess a histone N^{ε}-acetyl-lysine deacetylase activity, like yeast sir2 that was found in 2000 to catalyze a N^{ε}-acetyl-lysine deacetylation reaction on histone proteins.[2] In addition, sirtuin family members were also initially thought to possess a mono-ADP-ribosyltransferase activity.[3] However, as more research has been performed on the sirtuin family since 2000, especially during the past few years, we now know that histone proteins actually only constitute a minor group of the native sirtuin substrates.[4] A variety of nonhistone native sirtuin substrates have been identified in both prokaryotic and eukaryotic cells,[4,5] in the latter case, nonhistone sirtuin substrates are localized in nucleus (including nucleolus), mitochondrion, and cytosol (the three major intracellular compartments where the eukaryotic sirtuins also reside) and are involved in a variety of crucial cellular processes, such as gene transcription (in nucleus together with histone substrates) and metabolism (in mitochondrion and cytosol).[4,6–14] Also, many sirtuin family members are now known to possess more proficient activity of catalyzing the removal of N^{ε}-acyl-lysine's acyl groups bulkier than acetyl.[10,15–18] Typical examples of these bulkier acyl groups include crotonyl, malonyl, succinyl, glutaryl, 3-methylglutaryl, and myristoyl. Moreover, except for a type of sirtuins found in certain bacteria and fungi, the mono-ADP-ribosyltransferase activity for the majority of siruin family members is actually much weaker than the N^{ε}-acyl-lysine deacylase activity with the same sirtuin.[15] However, the sirtuin-catalyzed N^{ε}-acyl-lysine deacylation and mono-ADP-ribosyltransfer both have an absolute requirement of the redox coenzyme β-nicotinamide adenine dinucleotide (β-NAD^+) as the cosubstrate. Therefore, the sirtuin-catalyzed reactions play an important role in regulating a variety of crucial cellular processes. The sirtuin-catalyzed N^{ε}-acyl-lysine deacylation is further considered as a contemporary therapeutic target for human diseases, such as cancer, metabolic and neurodegenerative diseases.[19,20] In order to fully tap the therapeutic potential of this enzymatic deacylation, during the past few years we have seen a tremendous advancement in mechanistic elucidation, chemical modulator (inhibitor and activator) development, (chemical) biological and pharmacological exploration of the sirtuin-catalyzed deacylation. For this knowledge advancement, the employment of carefully designed chemical probes has played an elegant role, which will be delineated in the following sections according to the specific types of application for the developed chemical probes. It should be noted that the currently more useful chemical probes are the close structural analogs of N^{ε}-acyl-lysine.

2. CHEMICAL PROBES FOR THE MECHANISTIC STUDY OF THE SIRTUIN-CATALYZED DEACYLATION

Fig. 1A depicts the current version of the proposed chemical mechanism for the sirtuin-catalyzed deacylation. It should be noted that this mechanistic scheme would be consistent with the sequential kinetic mechanism for the sirtuin-catalyzed deacylation, that is, there is an obligatory formation of the ternary complex of a sirtuin with the lysine N^{ε}-acylated substrate and β-NAD^+ before the first chemical step (i.e., nicotinamide cleavage from β-NAD^+) occurs.[21] Moreover, this mechanistic scheme can be broken down into following stages: Stage 1, the critical formation of the covalent bond between the N^{ε}-acyl-lysine side chain amide oxygen and the C1′ of β-NAD^+, as highlighted in the depicted α-1′-O-alkylamidate intermediate (i.e., the intermediate I); Stage 2, the α-1′-O-alkylamidate intermediate I undergoes an intramolecular cyclization between the activated 2′-OH and the iminium carbon, affording the bicyclic intermediate (i.e., the intermediate II); Stage 3, the resolution of the bicyclic intermediate II in the presence of water affords the two end products of the sirtuin-catalyzed deacylation, that is, the deacylated product and 2′-O-acyl-ADP-ribose (2′-O-AADPR) that bears the acyl group originally on N^{ε}-acyl-lysine. The chemical change of this last stage is actually about how the dioxo ring of the bicyclic intermediate II gets opened up. To this end, what's shown in the bracket represents one possible route. The other three potential routes are shown in Fig. 1B. During the mechanistic interrogation of the sirtuin-catalyzed deacylation resulting in the fairly complex and impressive

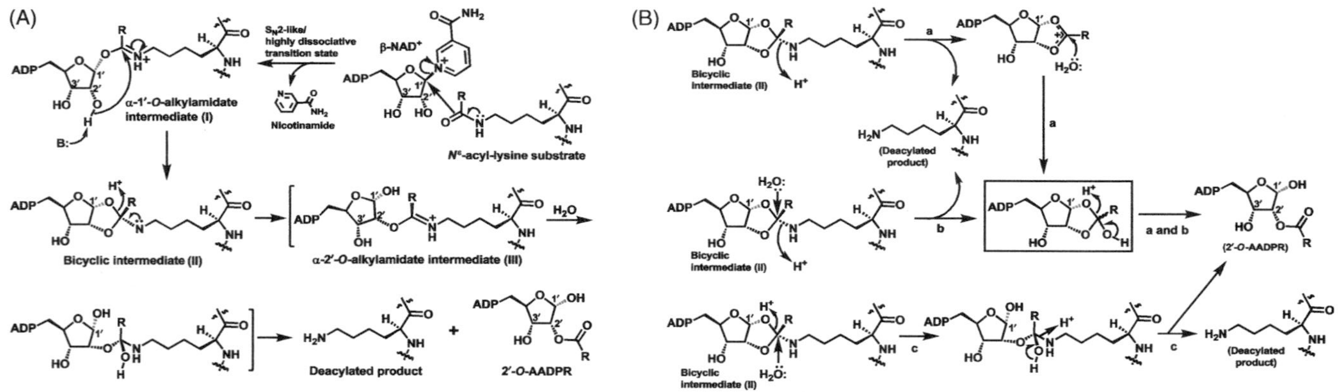

Fig. 1 (A) The current version of the proposed chemical mechanism for the sirtuin-catalyzed deacylation. Sample N^{ε}-acyl-lysine's acyl groups (R—(C=O)— include acetyl, crotonyl, malonyl, succinyl, glutaryl, 3-methylglutaryl, and myristoyl.[10,15–18] The stereochemistry at the tetrahedral carbon directly linked to the "R" group in the bicyclic intermediate II is inferred from the observed stereochemistry at the corresponding carbon in the α-1′-S-bicyclic intermediate shown in Fig. 4B (see below). *ADP*, adenosine diphosphate; B: refers to a general base, such as the neutral form of a histidine side chain.[22] (B) The three more potential routes accounting for the resolution of the bicyclic intermediate II in the presence of water, affording the two end products of the sirtuin-catalyzed deacylation, that is, the deacylated product and 2′-*O*-acyl-ADP-ribose (2′-*O*-AADPR). *Modified from Fig. 14 in Chen B, Zang W, Wang J, Huang Y, He Y, Yan L, Liu J, Zheng W. The chemical biology of sirtuins. Chem Soc Rev. 2015;44:5246–5264.*[10]

mechanistic scheme depicted in Fig. 1, carefully designed close structural analogs of N^{ϵ}-acyl-lysine have been employed as mechanistic probes, which is to be elaborated below.

To probe the covalent bond formation between the N^{ϵ}-acyl-lysine side chain amide oxygen and the C1′ of β-NAD^+ during the first stage of the sirtuin-catalyzed deacylation, the most direct evidence came from the experiments employing the two chemical probes shown in Fig. 2.

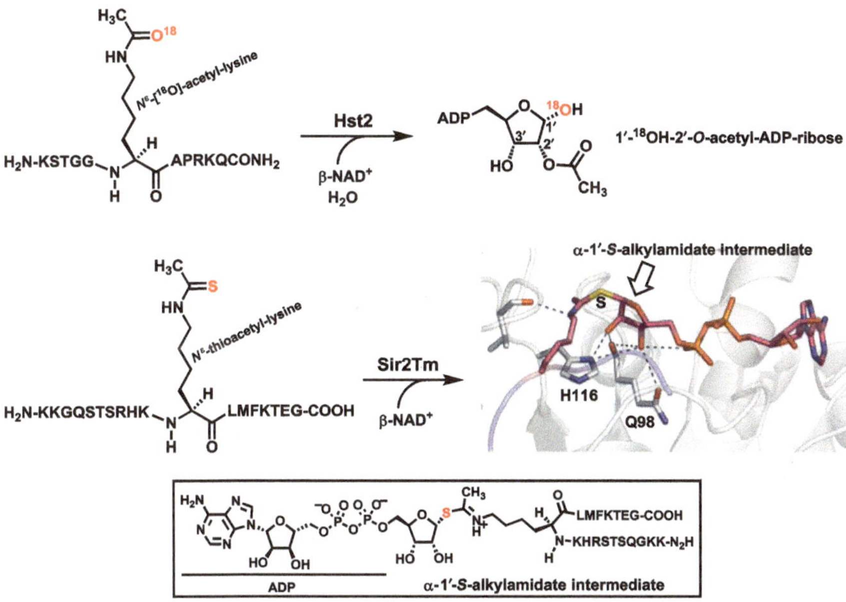

Fig. 2 The chemical structures of N^{ϵ}-[^{18}O]acetyl-lysine and N^{ϵ}-thioacetyl-lysine, the two chemical probes whose use, when incorporated into the depicted peptide sequences, has furnished the most direct evidence for the covalent bond formation between the N^{ϵ}-acyl-lysine side chain amide oxygen and the C1′ of β-NAD^+ during the first stage of the sirtuin-catalyzed deacylation. What are also shown are the chemical structure of 1′-^{18}OH-2′-*O*-acetyl-ADP-ribose, as well as the chemical structure (in the box) and the stick model of the α-1′-S-alkylamidate intermediate trapped within the Sir2Tm active site, the two key mechanistic species resulting from the sirtuin-catalyzed transformation on N^{ϵ}-[^{18}O]acetyl-lysine and N^{ϵ}-thioacetyl-lysine, respectively. *Hst2*, a yeast sirtuin; *Sir2Tm*, a bacterial sirtuin; *ADP*, adenosine diphosphate. *The close-up view showing the trapped α-1′-S-alkylamidate intermediate and its interactions with Sir2Tm active site amino acid residues was adapted with permission from Hawse WF, Hoff KG, Fatkins DG, Daines A, Zubkova OV, Schramm VL, Zheng W, Wolberger C. Structural insights into intermediate steps in the Sir2 deacetylation reaction. Structure. 2008;16: 1368–1377.*[23]

Specifically, when the histone H3 N-terminal tail peptide harboring the probe N^{ε}-[^{18}O]acetyl-lysine (i.e., H_2N-KSTGG-[N^{ε}-[^{18}O]acetyl-lysine]-APRKQCONH_2) was employed in an assay for the yeast Hst2-catalyzed deacetylation, the ^{18}O labeled 2′-*O*-AADPR (i.e., 1′-^{18}OH-2′-*O*-AADPR) was obtained.[22] This observation strongly suggests that there is a direct covalent bond formation between the N^{ε}-acyl-lysine side chain amide oxygen and the electrophilic C1′ of β-NAD^+ during normal sirtuin deacylation catalysis. Moreover, when soaking the crystal of the bacterial Sir2Tm bound to the N^{ε}-thioacetyl-lysine p53 peptide (i.e., H_2N-KKGQSTSRHK-[N^{ε}-thioacetyl-lysine]-LMFKTEG-COOH) with a β-NAD^+ cryoprotective solution, the α-1′-S-alkylamidate intermediate was found to be formed and trapped within the crystal, as revealed in the solved X-ray cocrystal structure (Protein Data Bank code: 3D81).[23] This and the similar observation of the stalled α-1′-S-alkylamidate intermediate with human SIRT3 and a N^{ε}-thioacetyl-lysine acetyl-CoA synthetase 2 peptide in a later structural study[24] also strongly suggest from a structure-based standpoint for a direct covalent bond formation between the N^{ε}-acyl-lysine side chain amide oxygen and the electrophilic C1′ of β-NAD^+ during normal sirtuin deacylation catalysis, with the formation of the α-1′-*O*-alkylamidate intermediate. It should be noted that a stalled α-1′-*S*-alkylamidate intermediate was also found to be formed with mass spectrometry detection in an assay of the yeast Hst2 with β-NAD^+ and a N^{ε}-thioacetyl-lysine histone H3 peptide.[25]

The interaction between the N^{ε}-acyl-lysine side chain amide oxygen and the electrophilic C1′ of β-NAD^+ is a nucleophilic substitution reaction, with the amide oxygen as the nucleophile, the C1′ of β-NAD^+ as the electrophile, and nicotinamide as the leaving group. Therefore, the next mechanistic question is whether this substitution reaction goes by the S_N1-like or the S_N2-like mechanism. To help to ascertain this, a few analogs of N^{ε}-acetyl-lysine substituted at its acetyl α-carbon with the strongly electron withdrawing fluorine atom (Fig. 3A) were each incorporated into position X in the following histone H3 N-terminal tail peptide sequence: H_2N-KSTGG-X-APRKQ-COOH, and the resulting peptides were employed in the assay with the yeast Hst2 together with the control peptide (X = N^{ε}-acetyl-lysine).[26] Based on the following findings in the study, it can be concluded that there is a nucleophilic participation of the side chain amide oxygen of N^{ε}-acetyl-lysine in the transition state for the sirtuin-catalyzed nicotinamide cleavage from β-NAD^+ (i.e., the reaction between the N^{ε}-acetyl-lysine side chain amide oxygen and the electrophilic C1′ of β-NAD^+) and a S_N2-like mechanism can be suggested. (1) The nicotinamide formation rate decreased

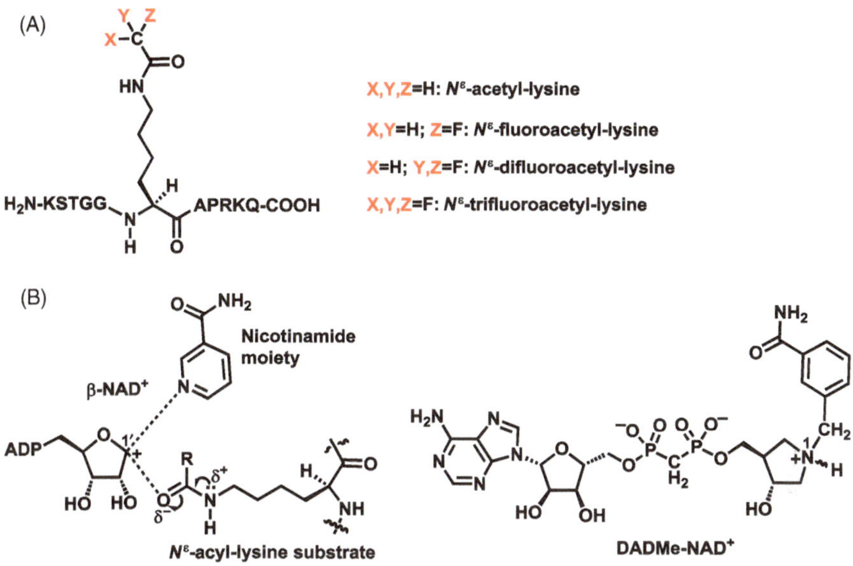

Fig. 3 (A) The peptides harboring N^{ε}-acetyl-lysine or its three fluorinated analogs employed in the assay with the yeast Hst2 to investigate if there is a nucleophilic participation of the side chain amide oxygen of N^{ε}-acetyl-lysine in the transition state for the sirtuin-catalyzed nicotinamide cleavage from β-NAD$^+$. (B) Left: A schematic illustration of the dissociative S_N2-like mechanism for the sirtuin-catalyzed nicotinamide cleavage from β-NAD$^+$, featuring a highly dissociative transition state with a strong oxacarbenium ion character and a mild nucleophilic participation from the side chain amide oxygen of N^{ε}-acyl-lysine. *ADP*, adenosine diphosphate. Because of this transition state structural feature, C1′ of β-NAD$^+$ is depicted to have a formal +1 charge. Right: The chemical structure of DADMe-NAD$^+$, a compound mimicking the nicotinamide-dissociated β-NAD$^+$. *Note*: To be consistent with the observed positioning of the N1 of DADMe-NAD$^+$ closer to N^{ε}-acetyl-lysine side chain amide oxygen in the 3-dimensional structure of the ternary complex of the bacterial Sir2Tm with a N^{ε}-acetyl-lysine peptide substrate and DADMe-NAD$^+$,[23] C1′ of β-NAD$^+$ is depicted closer to the side chain amide oxygen of N^{ε}-acyl-lysine in the transition state for the sirtuin-catalyzed nicotinamide cleavage from β-NAD$^+$.

(by ~1.8×10^3 to 6.1×10^5-fold) as the function of increasing number (from 1 to 3) of the strongly electron withdrawing fluorine atom, and thus of the decreased nucleophilicity of the side chain amide oxygen. (2) Per K_d measurement, the Hst2 active site was found to be also able to bind favorably to the fluorinated analogs of N^{ε}-acetyl-lysine, with K_d values of ~21, ~22, ~20, and ~3.3 μM for the histone H3 peptides respectively harboring N^{ε}-acetyl-lysine, N^{ε}-monofluoroacetyl-lysine, N^{ε}-difluoroacetyl-lysine, and N^{ε}-trifluoroacetyl-lysine.

However, as the amide oxygen is a weak nucleophile by itself, then how is a sirtuin able to accelerate its nucleophilic substitution reaction? It was suggested from a few later computational (e.g., ab initio QM/MM molecular dynamics simulation) and experimental (kinetic isotope effect measurement) studies with model sirtuins (e.g., bacterial Sir2Tm and archeal Sir2Af2) that the sirtuin-catalyzed nucleophilic substitution reaction between the N^{ε}-acyl-lysine side chain amide oxygen and the electrophilic C1′ of β-NAD^+ follows a concerted but dissociative S_N2-like mechanism involving a highly dissociative transition state featured with a strong oxacarbenium ion character yet a mild nucleophilic participation from N^{ε}-acyl-lysine's side chain amide oxygen (Fig. 3B).[27–29] Consistent with this transition state structural feature, the 3-dimensional structure of the ternary complex of the bacterial Sir2Tm with a N^{ε}-acetyl-lysine peptide substrate and DADMe-NAD^+ (a compound mimicking the nicotinamide-dissociated β-NAD^+,[30] Fig. 3B) (Protein Data Bank code: 3D4B) revealed that, while N^{ε}-acetyl-lysine and the nicotinamide mimic (i.e., benzamide) of DADMe-NAD^+ superimposed well with the respective positions of N^{ε}-acetyl-lysine and the nicotinamide moiety of β-NAD^+ in the structure of their Michaelis complex with Sir2Tm, the N1 of DADMe-NAD^+ was observed to be closer to N^{ε}-acetyl-lysine side chain amide oxygen.[23] Since N1 can be present as a trialkylammonium cation under physiological pH, this positioning of N1 in the ternary complex would also suggest that, in the transition state of the sirtuin-catalyzed nicotinamide cleavage from β-NAD^+, there is a movement of the C1′ of β-NAD^+ away from nicotinamide moiety and toward the side chain amide oxygen of N^{ε}-acyl-lysine, together with a concomitant migration of the positive charge on the nicotinamide N toward the C1′ of β-NAD^+, making it electrophilic enough to react with the weak nucleophile amide oxygen.

For the second stage of the sirtuin-catalyzed deacylation, the α-1′-*O*-alkylamidate intermediate I was proposed to be transformed into the bicyclic intermediate II, as depicted in Fig. 1A. The use of the two chemical probes shown in Fig. 4 generated the most direct experimental evidence for this proposed transformation.

Specifically, when a peptide harboring the close structural analog of N^{ε}-acetyl-lysine, that is, L-2-amino-7-carboxamidoheptanoic acid, was employed in the deacetylation assay with human SIRT1 or bacterial Sir2Tm, this analog was found to be processed by a sirtuin with the formation of the bicyclic intermediate (i.e., intermediate ii) shown in Fig. 4A that was long-lived enough to be detected using mass spectrometry.[31] As this bicyclic intermediate is a close structural analog of the bicyclic intermediate II

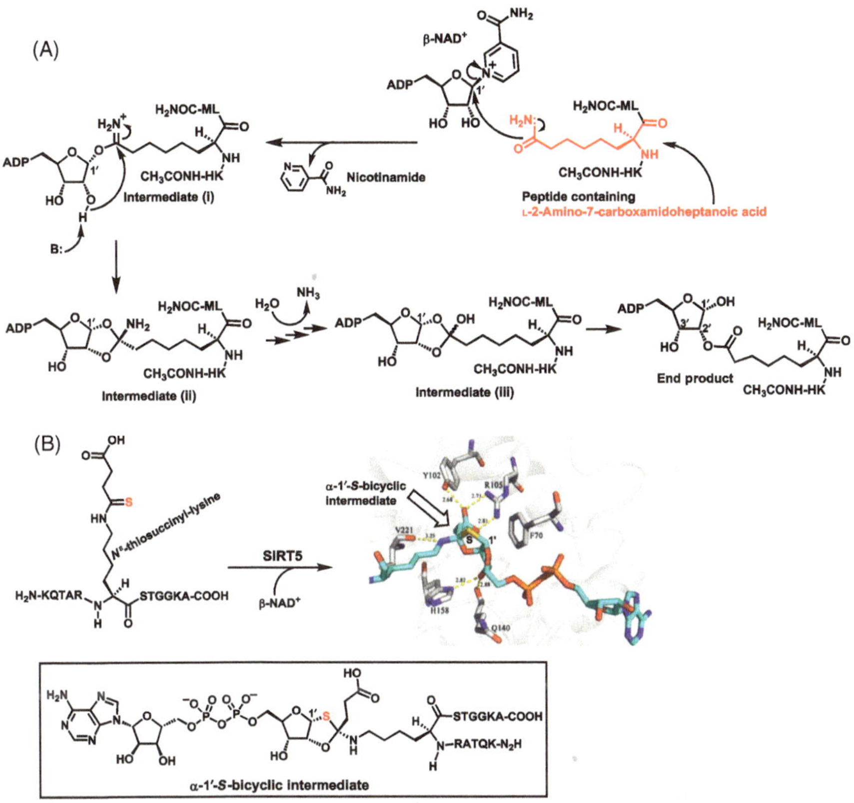

Fig. 4 (A) The proposed sirtuin processing of a p53 peptide containing the central L-2-amino-7-carboxamidoheptanoic acid residue *(shown in red)*. This peptide was found to be converted by human SIRT1 or bacterial Sir2Tm to the stalled bicyclic intermediate ii and (the stalled intermediate iii and/or the end product) which were detectable by mass spectrometry.[31] *ADP*, adenosine diphosphate; B: refers to a general base, such as the neutral form of a histidine side chain.[22] *Note*: The stereochemistry at the tetrahedral carbon directly linked to the "NH_2" group in the intermediate ii is inferred from the observed stereochemistry at the corresponding carbon in the α-1′-*S*-bicyclic intermediate shown in (B); whereas the stereochemistry at the tetrahedral carbon directly linked to the "OH" group in the intermediate iii is not defined. (B) A schematic illustration of using a histone H3 peptide harboring N^{ε}-thiosuccinyl-lysine as a chemical probe for the formation of the proposed bicyclic intermediate II during the normal sirtuin deacylation catalysis depicted in Fig. 1A. The α-1′-*S*-bicyclic intermediate (with the boxed chemical structure and the stick model) formed via the SIRT5-catalyzed processing of the probe is shown trapped at the SIRT5 active site. *Note*: The observed stereochemistry at the tetrahedral carbon directly linked to the "CH_2CH_2COOH" group in the α-1′-*S*-bicyclic intermediate is also indicated in the boxed chemical structure. *The close-up view showing this trapped intermediate and its interactions with SIRT5 active site amino acid residues was adapted with permission from Zhou Y, Zhang H, He B, Du J, Lin H, Cerione RA, Hao Q. The bicyclic intermediate structure provides insights into the desuccinylation mechanism of human sirtuin 5 (SIRT5). J Biol Chem. 2012;287:28307–28314.*[32]

in Fig. 1A, this observation argues strongly for the existence and the route of formation of the bicyclic intermediate II as depicted in Fig. 1A.

Moreover, when a binary complex of human SIRT5 and a histone H3 peptide harboring N^{ε}-thiosuccinyl-lysine was soaked with a cryoprotective solution of β-NAD^+, the α-1′-*S*-bicyclic intermediate shown in Fig. 4B was found to be formed (apparently from the SIRT5-catalyzed processing of the peptide) and trapped at the SIRT5 active site, as revealed by the solved X-ray cocrystal structure (Protein Data Bank code: 4F56).[32] This observation provides the first piece of the structure-based evidence arguing strongly for the existence and the route of formation of the bicyclic intermediate II as depicted in Fig. 1A.

As compared with the stalled bicyclic intermediate II depicted in Fig. 4A, the stalled α-1′-*S*-bicyclic intermediate in Fig. 4B is structurally closer to the bicyclic intermediate II depicted in Fig. 1A proposed for the sirtuin deacylation catalysis; however, both findings collectively argue for the chemical mechanism conservation for the deacylation reaction catalyzed by different sirtuin family members. As for the case with the above SIRT5 crystallographic study, even though N^{ε}-thiosuccinyl-lysine is simply the thio version of N^{ε}-succinyl-lysine, just like N^{ε}-thioacetyl-lysine versus N^{ε}-acetyl-lysine, the SIRT5-catalyzed processing of N^{ε}-thiosuccinyl-lysine led to the stalling of the α-1′-*S*-bicyclic intermediate; whereas the Sir2Tm (or SIRT3, Hst2)-catalyzed processing of N^{ε}-thioacetyl-lysine led to the stalling of the α-1′-*S*-alkylamidate intermediate. Therefore, active site variation also exists among different sirtuin family members.

As for the resolution of the bicyclic intermediate II in the presence of water to afford the two end products during the third stage of the sirtuin-catalyzed deacylation, the route depicted in Fig. 1A represents one possibility, which has been proposed based on the trapping of the thio version of the α-2′-*O*-alkylamidate intermediate III (i.e., 1′-SH-α-2′-*O*-alkylamidate intermediate) within the SIRT2 active site, as revealed by the solved cocrystal structure (Protein Data Bank code: 4X3O), following the incubation of human SIRT2 with β-NAD^+ and a peptide harboring N^{ε}-thiomyristoyl-lysine (Fig. 5).[33] However, in our opinion, this proposal needs to be treated with caution in terms of its applicability to the sirtuin family members as a whole. Specifically, since thiolate ($-S^-$) is a better leaving group than alkoxide ($-O^-$), C—S bond can be cleaved in the absence of a general acid while abstracting a proton from bulk reaction solution; however, C—O bond cleavage would be more possible with the help of a general acid. With regard to this, it would be essential to see if there is any such general acid

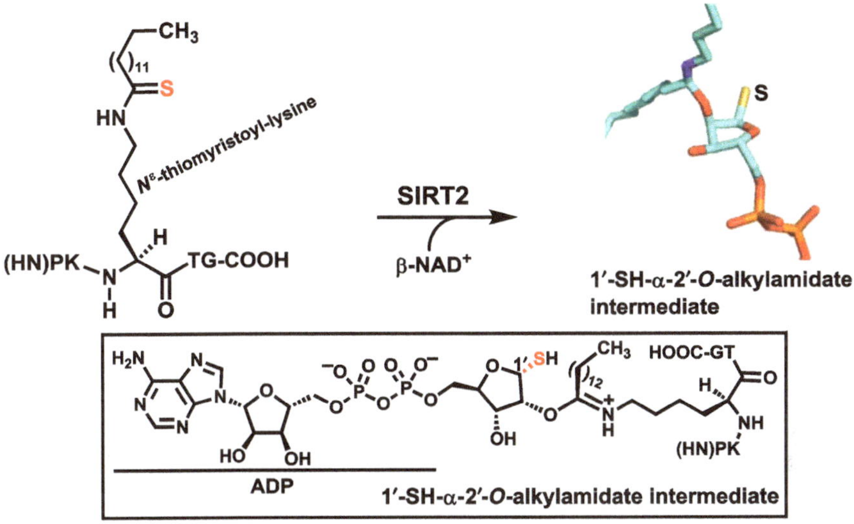

Fig. 5 A schematic illustration of the SIRT2-catalyzed transformation of the N^{ε}-thiomyristoyl-lysine-containing TNFα peptide into the stalled 1′-SH-α-2′-*O*-alkylamidate intermediate trapped within the SIRT2 active site. The chemical structure (in the box) and the stick model of the trapped 1′-SH-α-2′-*O*-alkylamidate intermediate are shown. "HN" denotes Pro's free α-NH group. *The stick model of the trapped 1′-SH-α-2′-O-alkylamidate intermediate was adapted with permission from Wang Y, Fung YM, Zhang W, He B, Chung MW, Jin J, Hu J, Lin H, Hao Q. Deacylation mechanism by SIRT2 revealed in the 1′-SH-2′-O-myristoyl intermediate structure. Cell Chem Biol. 2017;24:339–345.*[33]

appropriately situated in the solved cocrystal SIRT2 structure. If there is no such general acid, then the observed formation of the 1′-SH-α-2′-*O*-alkylamidate intermediate may be unique to the use of N^{ε}-thiomyristoyl-lysine; consistent with this reasoning, the SIRT2 (or sirtuin in general) deacylation may not go through the α-2′-*O*-alkylamidate intermediate III and the downstream catalytic intermediate with the N^{ε}-acyl-lysine substrate as depicted in Fig. 1A.

Fig. 1B depicts the three more routes potentially responsible for the collapse of the bicyclic intermediate II in the presence of water to afford the two end products. The aforementioned study with the N^{ε}-acetyl-lysine analog L-2-amino-7-carboxamidoheptanoic acid and SIRT1 or Sir2Tm revealed not only the formation of the longer-lived bicyclic intermediate II shown in Fig. 4A, but also the formation of one or more species whose exact mass corresponds to the longer-lived hemiorthoester intermediate III and/or the end product also shown in Fig. 4A.[31] It should be noted that this hemiorthoester intermediate mimics the hemiorthoester intermediate (the boxed structure) shared by pathways A and B depicted in Fig. 1B, therefore, at least for SIRT1 or Sir2Tm, pathway A or B could be the pathway responsible for the collapse of the bicyclic intermediate II in the presence of water. Obviously, more work still needs to be done to tease out the exact pathway or pathways responsible for the third stage of the sirtuin-catalyzed deacylation. One possible scenario is that different sirtuin family members might employ different pathways.

3. ACTIVITY-BASED PROBES FOR THE SIRTUIN-CATALYZED DEACYLATION

Since sirtuin family members are intracellular enzymes whose enzymatic reactions play an important regulatory role in crucial cellular processes, a real-time monitoring of their enzymatic activities inside cells would be important for enhancing our understanding of the inner workings of the sirtuin-mediated cellular signaling networks and their pathological roles in human diseases as well.

While the expression (mRNA and protein) levels of sirtuin family members can be determined with Northern blot and Western blot, respectively, it is the catalytic activities of sirtuins that primarily underlie their (patho) physiological functions; therefore, the reporting of sirtuins' catalytic activities within the cellular context is critical to enhancing our understanding of the functioning roles of sirtuin proteins as enzymes. To this end, we have already seen a few studies on the activity-based sirtuin protein profiling.

Inspired by the above-described ability of a sirtuin deacetylase activity to catalyze the formation of the stalled α-1′-*S*-alkylamidate intermediate from a N^{ε}-thioacetyl-lysine substrate and β-NAD^+ (see Fig. 2), a recent study[34] reported the use of a β-NAD^+ analog carrying an affinity tag in the assay with a N^{ε}-thioacetyl-lysine substrate and a recombinant sirtuin so as to pull down (via the affinity tag on the β-NAD^+ analog) the enzymatically active sirtuin deacetylase with the stalled α-1′-*S*-alkylamidate intermediate bound at its active site. This is the first report of detecting and isolating the enzymatically active sirtuin deacetylases, however, this strategy is limited by the lack of a covalent linkage between the stalled α-1′-*S*-alkylamidate intermediate and a sirtuin protein.

A few fluorescent probes for detecting the deacylase (including deacetylase) activities of recombinant sirtuins or the sirtuins in cell lysates (or inside cells) have been developed. Fig. 6 depicts these probes and how they work. The key feature of the working principle of all of these probes is to rely on the sirtuin-catalyzed deacylation to turn on fluorescence or to relieve the quenching [via fluorescence resonance energy transfer (FRET)] of the fluorescence of a fluorophore, so that the deacylase activity of a sirtuin can be correlated with the degree of fluorescence increase following the deacylation of a probe by the sirtuin. It should be noted that the genetically encoded protein-based probe D can be employed to detect the sirtuin deacylase activity inside cells, as this probe is to be generated inside cells via unnatural mutagenesis making use of the gene expression machinery of a host cell. In addition, it is interesting that the peptidic probes B and C-2 were shown to be proteolytically stable enough and cell permeable so that they were able to detect the intracellular sirtuin deacylase activity. It should also be noted that, while the use of these fluorescent probes entails a highly sensitive detection of sirtuin deacylase activity, these probes have either a low or an unknown substrate selectivity among different sirtuin family members (e.g., among the seven mammalian sirtuins) and versus the deacylases outside of the sirtuin family [e.g., the Zn^{2+}-containing histone deacetylase (HDAC) family[35]], nevertheless, these current fluorescent probes are valuable lead compounds for developing future superior fluorescent probes with superior sirtuin substrate selectivity. If these future probes are cell permeable, their use would enable a better assessment of the deacylase activity of each individual sirtuin inside cells. To enhance the proteolytic stability and cell permeability, peptidomimetic- or cyclic peptide-based version of the linear peptidic probes would be worth of pursuing.[36]

In addition to the above-described studies, chemical probes have also been developed for the identification of the sirtuin(s) responsible for the recognition of a particular posttranslational lysine N^{ε}-acylation. Fig. 7 depicts these probes and how they work. Specifically, all of these probes are photo-affinity labeling agents (or just photo-affinity labels), hoping that a covalent bond between the probe and a sirtuin will be formed following the activation by UV irradiation of the photo-reactive group in the probe whose N^{ε}-acyl-lysine has been transiently bound at a sirtuin active site. All of these photo-affinity labels have a tripartite structural makeup, consisting of a N^{ε}-acyl-lysine core installed with appropriately placed photo-reactive group (e.g., benzophenone and diazirine) and terminal alkyne (a bioorthogonal handle for the selective detection/isolation of the captured sirtuin(s) for the subsequent sirtuin identification via quantitative proteomics). Of note, sirtuin detection can be realized with in-gel fluorescence, and sirtuin isolation

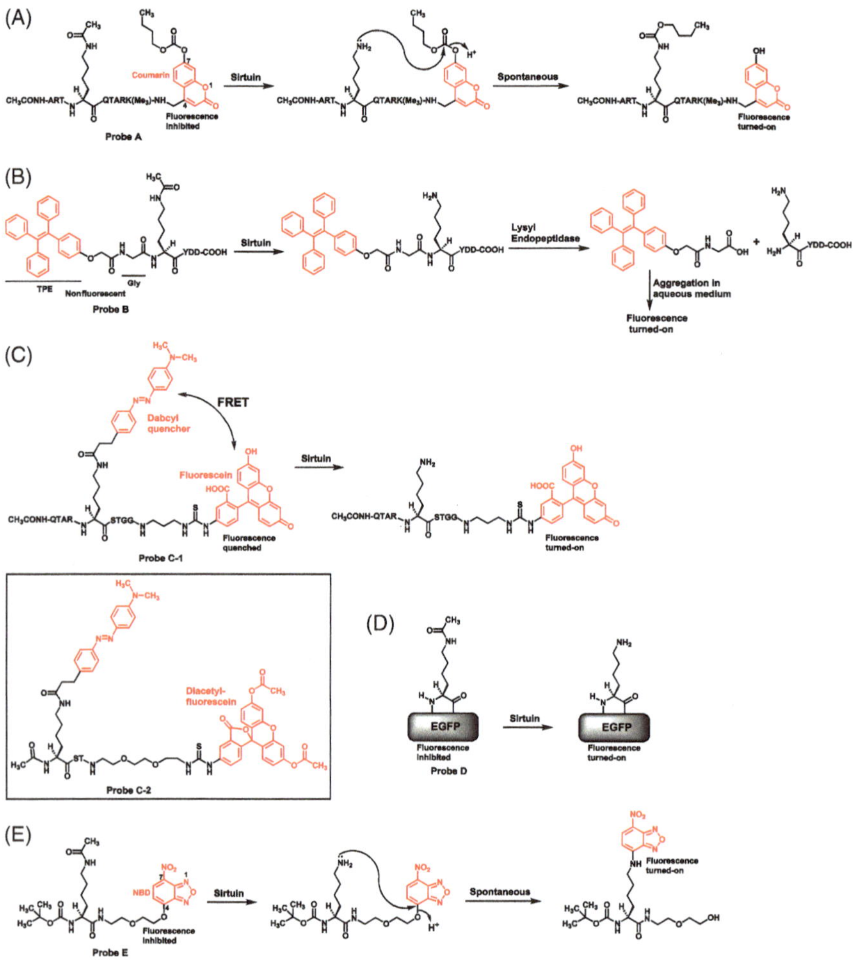

Fig. 6 The fluorescent probes that have been developed for detecting the deacylase (including deacetylase) activities of sirtuins. (A) The coumarin-based probe A[37] that is composed of a N^{ε}-acetyl-lysine sirtuin substrate peptide (derived from the N-terminal tail of histone H3 protein) connected to a coumarin derivative, whose 7-OH is acylated in the form of a carbonate ester so that the coumarin fluorescence is inhibited. However, following the sirtuin-catalyzed N^{ε}-acetyl-lysine deacetylation of the probe and the ensuing spontaneous intramolecular reaction as depicted with the liberated free 7-OH, the coumarin fluorescence is turned on. The chemical structure of coumarin is colored in *red*. (B) The tetraphenylethene (TPE)-based probe B[38] that is composed of a hydrophilic N^{ε}-acetyl-lysine SIRT1 substrate peptide [derived from the N-terminal domain of liver kinase B1 (LKB1)] connected to the hydrophobic TPE (*colored in red*) is nonfluorescent. However, following the SIRT1-catalyzed N^{ε}-acetyl-lysine deacetylation of the probe and the ensuing proteolysis by lysyl endopeptidase, the liberated TPE-Gly-COOH would aggregate in aqueous solution resulting in the turned-on fluorescence

can be performed via the biotin–streptavidin specific pull-down. It should be noted that, since the sirtuin-catalyzed deacylation reaction was suggested to obey an ordered sequential kinetic mechanism in which the N^{ε}-acyl-lysine substrate binds to a sirtuin before β-NAD^+ binds,[21] the binding of these photo-affinity labels to sirtuin(s) in the absence of β-NAD^+ would presumably represent a productive binding of the N^{ε}-acyl-lysine substrate.

Among the photo-affinity labels depicted in Fig. 7, those based on N^{ε}-succinyl-lysine[42] and N^{ε}-malonyl-lysine[43] are able to label SIRT5, a sirtuin with proficient desuccinylase/demalonylase activities. Those based on N^{ε}-acetyl-lysine[41,43] are able to label SIRT1/2/3, sirtuins with proficient deacetylase activity. Moreover, the use of that based on N^{ε}-myristoyl-lysine revealed that it was able to more proficiently label SIRT2 than SIRT6,[44]

◀ with the aggregation-induced emission characteristics. (C) A FRET-based probe C-1[39] that is composed of a N^{ε}-acyl-lysine sirtuin substrate peptide (a nonapeptide derived from the N-terminal tail of histone H3 protein) connected to the fluorophore fluorescein (*colored in red*). Since the acyl group of N^{ε}-acyl-lysine is a derivative of the fluorophore 4-(4-dimethylaminophenylazo)benzoyl (dabcyl) *(also colored in red)*, the fluorescence of fluorescein would be quenched by this dabcyl-type quencher via FRET. However, following the sirtuin-catalyzed N^{ε}-acyl-lysine deacylation of the probe with the removal of the dabcyl-type quencher, the fluorescein fluorescence is turned on. This strategy works because of the recent demonstration that multiple sirtuins (e.g., SIRT1/2/3/6) can proficiently catalyze the fatty-acyl removal from the N^{ε}-fatty-acylated lysine.[10,15] The probe C-2[39] is a structurally simplified derivative of probe C-1 (tripeptidic vs. nonapeptidic) and was shown to be able to detect the SIRT1 deacylase activity inside cells due to its proteolytic stability and cell permeability. Of note, the more hydrophobic and thus presumably more cell permeable derivative of fluorescein (i.e., diacetylfluorescein) was used in probe C-2. Being a "prodrug" form of fluorescein, diacetylfluorescein ought to be converted back to fluorescein by the intracellular esterases. (D) Probe D[40] is a protein-based probe in which the enhanced green fluorescent protein (EGFP) was engineered by unnatural mutagenesis to be equipped with the N^{ε}-acetyl-lysine residue at position 85 in place of the native lysine residue. Since this lysine residue is essential for the fluorophore maturation, the EGFP bearing the N^{ε}-acetyl-lysine residue at its position 85 would lead to a fluorescence inhibition. However, since this N^{ε}-acetyl-lysine-containing EGFP is a good substrate for multiple sirtuins (e.g., SIRT1/2/3/5), following the sirtuin-catalyzed N^{ε}-acetyl-lysine deacetylation, the EGFP fluorescence would be turned on. (E) A 7-nitro-2,1,3-benzoxadiazole (NBD)-based probe E[41] that is composed of a N^{α}-Boc-N^{ε}-acetyl-lysine sirtuin substrate connected to a NBD derivative whose C-4 is substituted with the strongly electron withdrawing oxygen so that the NBD fluorescence is inhibited. However, following the sirtuin-catalyzed N^{ε}-acetyl-lysine deacetylation of the probe and the ensuing spontaneous intramolecular reaction as depicted with the generation of the C-4 nitrogen substituted NBD derivative, the NBD fluorescence is turned on. The chemical structure of NBD is colored in *red*.

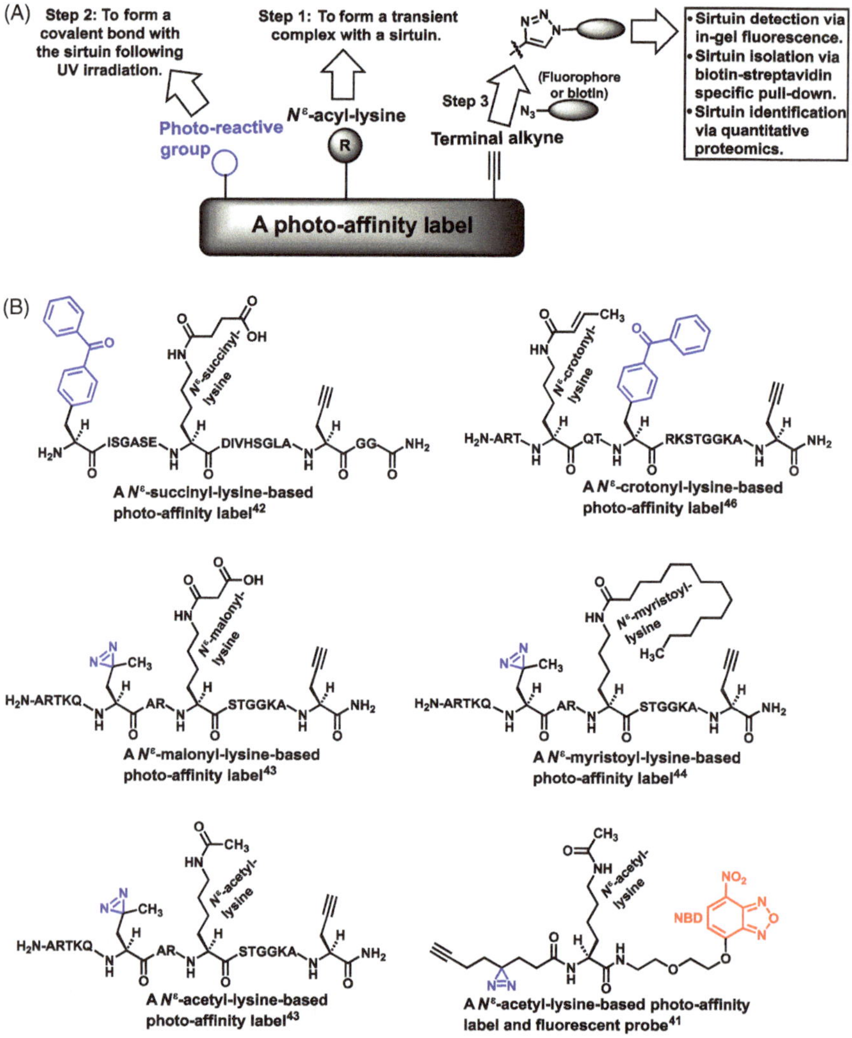

Fig. 7 (A) A schematic illustration of the photo-affinity labeling for the subsequent sirtuin detection, isolation, and identification. (B) The photo-affinity labels that have been developed to label the sirtuins recognizing the corresponding N^{ε}-acyl-lysine. The photo-reactive group in each label is colored in *blue*, and the fluorophore NBD in the last label is colored in *red*.

suggesting that SIRT2 also possesses a robust demyristoylase activity. This finding is also consistent with the previous in vitro findings.[45] Further, when the photo-affinity label based on N^{ε}-crotonyl-lysine was employed in combination with the stable isotope labeling of amino acids in cell culture

(SILAC) technology, SIRT3 was identified as an endogenous N^{ε}-crotonyl-lysine decrotonylase.[46]

As a potential limitation of using the photo-affinity labels depicted in Fig. 7, besides sirtuins they could also simultaneously label the deacylases outside of the sirtuin family (e.g., the Zn^{2+}-containing HDAC family[35]) while being unable to differentiate between these different types of the deacylases by using these labels per se. However, if this situation arises, these deacylases from different families could be differentiated by taking advantage of their different biochemical characteristics. For example, while the sirtuin deacylase activity is sensitive to the inhibition by the pan-inhibitor nicotinamide, the Zn^{2+}-containing HDACs are all sensitive to the inhibition by their pan-inhibitor trichostatin A.[35] Second, the photo-affinity labels depicted in Fig. 7 could also simultaneously label the N^{ε}-acyl-lysine recognizing protein domains (or readers, e.g., the N^{ε}-acetyl-lysine recognizing bromodomains[47]) yet could also be unable to differentiate these two types of target proteins by using these labels per se, except for the last label (see below). While further biochemical assessment could differentiate them taking advantage of their different biochemical characteristics, the last photo-affinity label depicted in Fig. 7 could realize the differentiation by itself. Specifically, even though this reagent is able to simultaneously label and pull-down sirtuins and readers, it is able to differentiate them since only the former can catalyze the N^{ε}-acetyl-lysine deacetylation of the reagent, which will trigger a spontaneous intramolecular reaction turning on the NBD fluorescence, a process same as that depicted above in Fig. 6E.

4. CHEMICAL PROBES FOR THE IDENTIFICATION OF NATIVE SUBSTRATES FOR THE SIRTUIN-CATALYZED DEACYLATION

The identification of the native substrates for the sirtuin-catalyzed deacylation would enhance our functional understanding of this enzymatic reaction. To this end, a quantitative proteomics approach can be taken, as exemplified by a recent study of identifying potential native SIRT7 deacetylation substrates via a SILAC-based quantitative proteomics approach;[48] an enhanced understanding of the substrate recognition behavior at sirtuin active site would also help to identify the native sirtuin deacylation (including deacetylation) substrates. However, a chemical biological approach to such substrate identification would also be desirable.

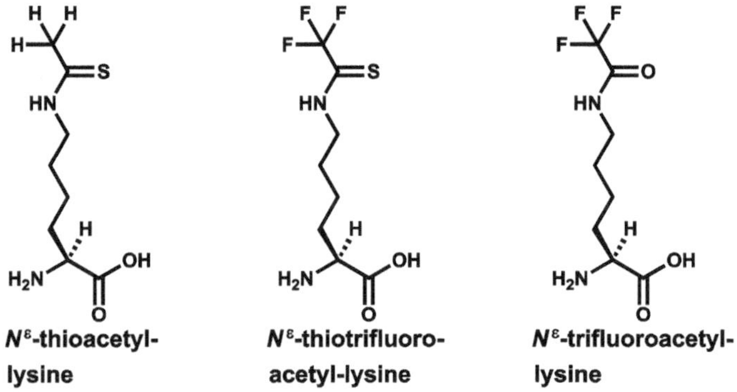

Fig. 8 A structural comparison of $N^ε$-thiotrifluoroacetyl-lysine with $N^ε$-thioacetyl-lysine and $N^ε$-trifluoroacetyl-lysine.

In a recent study with human SIRT3 as the model sirtuin,[49] the authors constructed on the cellulose membrane combinatorial 9-amino acid peptide libraries whose members all harbor $N^ε$-thiotrifluoroacetyl-lysine (structure depicted in Fig. 8) as the central residue, screened for high-affinity SIRT3 binding peptides, and subsequently subjected the resulting experimental data to machine learning analysis to establish binding trends and to make a SIRT3binding affinity prediction for all the lysine sites in the entire proteome of mitochondria where mature SIRT3 resides. Amazingly, the in-solution kinetic validation in the study revealed that the predicted SIRT3binding affinities for a set of 24 $N^ε$-acetyl-lysine peptides correlated with the k_{cat}/K_m values for their deacetylation catalyzed by SIRT3. By using the mitochondrial SIRT3 as a model sirtuin, this study has not only established an unbiased screening strategy, but also discovered multiple potential native substrates for SIRT3 in the mitochondrial proteome.

It should be noted that the key to this study is the use of $N^ε$-thiotrifluoroacetyl-lysine which can be regarded as a chimera of $N^ε$-thioacetyl-lysine and $N^ε$-trifluoroacetyl-lysine (see Fig. 8 for a structural comparison of these three $N^ε$-acetyl-lysine analogs). Therefore, like $N^ε$-trifluoroacetyl-lysine and $N^ε$-thioacetyl-lysine that were previously found by the same research group to bind tighter than $N^ε$-acetyl-lysine to yeast Hst2 via the K_d measurement for their respective histone H3 peptides [K_d (μM): ~3.3, ~4.7, ~21, respectively,[25,26] also see above], $N^ε$-thiotrifluoroacetyl-lysine would also bind tightly to sirtuin active site. In fact, the K_d value of the corresponding histone H3 peptide harboring $N^ε$-thiotrifluoroacetyl-lysine to yeast Hst2 was determined to be ~1 μM in this study. Moreover, like

N^{ε}-trifluoroacetyl-lysine whose incorporation in a histone H3 peptide sequence in place of N^{ε}-acetyl-lysine was found previously to decrease the rate of the yeast Hst2-catalzyed nicotinamide formation by $\sim 6.1 \times 10^5$-fold,[26] also as above described, N^{ε}-thiotrifluoroacetyl-lysine would also be only very weakly support the sirtuin-catalyzed nicotinamide cleavage from β-NAD^+ (the first chemical step of the sirtuin deacylation catalysis, see above).

5. CHEMICAL PROBES USED AS THE CATALYTIC MECHANISM-BASED INHIBITORY WARHEADS FOR THE SIRTUIN-CATALYZED DEACYLATION

Given the ever-increasingly demonstrated significant regulatory role of sirtuins in various vital cellular processes, such as gene transcription, metabolism, and DNA damage repair,[6–9] the sirtuin deacylation reaction has also been regarded as a contemporary therapeutic target for human diseases, such as cancer, neurodegenerative and metabolic diseases.[19,20] Therefore, chemical modulators (inhibitors and activators) of the sirtuin deacylation have been actively pursued during the past few years (see ref. 50 and Chapter 2 in this volume for latest accounts). Here we would like to highlight the amazing power of some chemical probes for sirtuin mechanistic research as highly efficient catalytic mechanism-based inhibitory warheads for the sirtuin-catalyzed deacylation.

N^{ε}-thioacyl-lysines constitute the first class of the catalytic mechanism-based inhibitory warheads for the sirtuin-catalyzed deacylation reaction, with N^{ε}-thioacetyl-lysine being the first N^{ε}-thioacyl-lysine discovered during 2006–07.[25,51] As described above, the use of N^{ε}-thioacyl-lysines in place of N^{ε}-acyl-lysines in sirtuin deacylation substrates leads to the formation of the stalled thio version of the catalytic intermediates I, II, and III (see Figs. 1A, 2, 4B, and 5). Each of these stalled intermediates could behave as a tight binding bisubstrate analog inhibitor for the sirtuin deacylation reaction in that its bipartite structural makeup is composed of two parts respectively derived from N^{ε}-thioacyl-lysine and β-NAD^+, and its formation via sirtuin catalysis could be regarded as a perfect way of constructing a bisubstrate analog sirtuin inhibitor.

The above-described L-2-amino-7-carboxamidoheptanoic acid represents the prototype of the second class of catalytic mechanism-based inhibitory warheads for the sirtuin-catalyzed deacylation reaction. Its processing by sirtuin has been shown to lead to the formation of one or more stalled catalytic intermediates (see Fig. 4A) which can be also regarded as the tight-binding bisubstrate analog sirtuin inhibitors.

One attractive feature of these catalytic mechanism-based sirtuin inhibitory warheads is that, by varying the terminal thioacyl (for N^{ε}-thioacyl-lysines) or N-substituent (for L-2-amino-7-carboxamidoheptanoic acid), the corresponding warheads would be selective toward different sirtuin family members, since these warheads would be first recognized as substrates for the sirtuin deacylation reaction; therefore, the acyl head specificity of the sirtuin N^{ε}-acyl-lysine substrate would also be applicable to these warheads. Within this context, it is currently known that, among the seven mammalian sirtuins (i.e., SIRT1–7), SIRT1/2/3/6 can proficiently catalyze the defatty-acylation of N^{ε}-fatty-acyl-lysine (e.g., N^{ε}-myristoyl-lysine), SIRT1/2/3 also possess a robust N^{ε}-acetyl-lysine deacetylase activity, and SIRT3 also is a native N^{ε}-crotonyl-lysine decrotonylase[10,15]; SIRT5 can proficiently catalyze the demalonylation/desuccinylation/deglutarylation of N^{ε}-malonyl-/succinyl-/glutaryl-lysine, respectively[10,15]; SIRT4 can proficiently catalyze the removal of the glutaryl and 3-methylglutaryl groups from N^{ε}-glutaryl-lysine and N^{ε}-(3-methylglutaryl)-lysine, respectively[16]; SIRT7 was found to proficiently catalyze the desuccinylation of N^{ε}-succinyl-lysine.[18] Moreover, SIRT7 was also found to possess appreciable deacetylase and defatty-acylase (e.g., demyristoylase) activities in the presence of dsDNA, rRNA, or tRNA.[17,52] Therefore, together with the exceptional sirtuin deacylase inhibitory power of these warheads, the family of the inhibitors endowed with them hold a great promise for ultimately developing potential therapeutic agents for human diseases.

6. CONCLUSIONS AND PERSPECTIVES

In conclusion, this chapter has furnished an updated account on the tremendous help that carefully designed chemical probes have lent to the studies on the N^{ε}-acyl-lysine deacylation, the primary type of reaction catalyzed by sirtuin family members. Specifically, chemical probes have been developed for: (1) a more direct dissection of the chemical mechanism, (2) reporting the intracellular activity, and (3) identifying the native substrates of the sirtuin-catalyzed deacylation. Moreover, chemical probes have also been developed for the detection, isolation, and identification of the sirtuin(s) responsible for the recognition of a particular posttranslational lysine N^{ε}-acylation. Intriguingly, some chemical probes (especially N^{ε}-thioacyl-lysines) for sirtuin mechanistic studies have also been found to be extremely powerful catalytic mechanism-based inhibitory warheads for the sirtuin-catalyzed deacylation. It would be rewarding that novel exquisite chemical probes also be developed in the future to tackle further sirtuin mechanistic and functional questions. Concerning the sirtuin-catalyzed N^{ϵ}-acyl-lysine deacylation, a further dissection of how the bicyclic intermediate II depicted in Fig. 1 is resolved in the presence of water to afford the deacylated product and 2′-*O*-AADPR and how to assess the regulatory role of the physicochemically variable and intrinsically disordered N-/C-termini of sirtuin family members[53] from a chemical biology standpoint are the two areas in which the future chemical probes may be entitled to a marvelous use. Despite a great success of the above-described N^{ϵ}-thiotrifluoroacetyl-lysine in helping to identify the native substrates for the model sirtuin SIRT3 in the mitochondrial proteome, to more directly identify the native substrates of the sirtuin-catalyzed deacylation reaction from the chemical biology standpoint would still be desirable, and as such devising novel chemical probes would be rewarding. One key characteristic for the future sirtuin chemical probes intended for the in vivo studies would be an enhanced selectivity among different sirtuin family members and versus other protein N^{ϵ}-acyl-lysine deacylases (e.g., the Zn^{2+}-dependent family of deacylases) and other intracellular proteins also able to recognize N^{ϵ}-acyl-lysines (e.g., the N^{ϵ}-acetyl-lysine recognizing bromodomains).

ACKNOWLEDGMENTS

The work on sirtuins in W.Z.'s laboratory has been supported by the following: National Natural Science Foundation of China (Grant No. 21272094), the Jiangsu Provincial Specially Appointed Professorship, the Jiangsu Provincial "Innovation and Venture Talents" award plan, Jiangsu University, the US National Institutes of Health (CA152972), the James L. and Martha J. Foght Endowment, and University of Akron.

REFERENCES

1. Greiss S, Gartner A. Sirtuin/Sir2 phylogeny, evolutionary considerations and structural conservation. *Mol Cells.* 2009;28:407–415.
2. Imai S, Armstrong CM, Kaeberlein M, Guarente L. Transcriptional silencing and longevity protein Sir2 is an NAD-dependent histone deacetylase. *Nature.* 2000;403:795–800.
3. Frye RA. Characterization of five human cDNAs with homology to the yeast SIR2 gene: Sir2-like proteins (sirtuins) metabolize NAD and may have protein ADP-ribosyltransferase activity. *Biochem Biophys Res Commun.* 1999;260:273–279.
4. Martínez-Redondo P, Vaquero A. The diversity of histone versus nonhistone sirtuin substrates. *Genes Cancer.* 2013;4:148–163.

5. Carabetta VJ, Cristea IM. Regulation, function, and detection of protein acetylation in bacteria. *J Bacteriol.* 2017;199(16):e00107–e00117.
6. Elkhwanky MS, Hakkola J. Extranuclear sirtuins and metabolic stress. *Antioxid Redox Signal.* 2017.
7. Dai Y, Faller DV. Transcription regulation by class III histone deacetylases (HDACs)-sirtuins. *Transl Oncogenomics.* 2008;3:53–65.
8. Choi JE, Mostoslavsky R. Sirtuins, metabolism, and DNA repair. *Curr Opin Genet Dev.* 2014;26:24–32.
9. Sebastián C, Mostoslavsky R. The role of mammalian sirtuins in cancer metabolism. *Semin Cell Dev Biol.* 2015;43:33–42.
10. Chen B, Zang W, Wang J, Huang Y, He Y, Yan L, Liu J, Zheng W. The chemical biology of sirtuins. *Chem Soc Rev.* 2015;44:5246–5264.
11. Chen S, Seiler J, Santiago-Reichelt M, Felbel K, Grummt I, Voit R. Repression of RNA polymerase I upon stress is caused by inhibition of RNA-dependent deacetylation of PAF53 by SIRT7. *Mol Cell.* 2013;52:303–313.
12. Zhang PY, Li G, Deng ZJ, Liu LY, Chen L, Tang JZ, Wang YQ, Cao ST, Fang YX, Wen F, Xu Y, Chen X, Shi KQ, Li WF, Xie C, Tang KF. Dicer interacts with SIRT7 and regulates H3K18 deacetylation in response to DNA damaging agents. *Nucleic Acids Res.* 2016;44:3629–3642.
13. Kiran S, Chatterjee N, Singh S, Kaul SC, Wadhwa R, Ramakrishna G. Intracellular distribution of human SIRT7 and mapping of the nuclear/nucleolar localization signal. *FEBS J.* 2013;280:3451–3466.
14. Matsushita N, Yonashiro R, Ogata Y, Sugiura A, Nagashima S, Fukuda T, Inatome R, Yanagi S. Distinct regulation of mitochondrial localization and stability of two human Sirt5 isoforms. *Genes Cells.* 2011;16:190–202.
15. Bheda P, Jing H, Wolberger C, Lin H. The substrate specificity of sirtuins. *Annu Rev Biochem.* 2016;85:405–429.
16. Anderson KA, Huynh FK, Fisher-Wellman K, Stuart JD, Peterson BS, Douros JD, Wagner GR, Thompson JW, Madsen AS, Green MF, Sivley RM, Ilkayeva OR, Stevens RD, Backos DS, Capra JA, Olsen CA, Campbell JE, Muoio DM, Grimsrud PA, Hirschey MD. SIRT4 is a lysine deacylase that controls leucine metabolism and insulin secretion. *Cell Metab.* 2017;25:838–855.
17. Tong Z, Wang M, Wang Y, Kim DD, Grenier JK, Cao J, Sadhukhan S, Hao Q, Lin H. SIRT7 is an RNA-activated protein lysine deacylase. *ACS Chem Biol.* 2017;12:300–310.
18. Li L, Shi L, Yang S, Yan R, Zhang D, Yang J, He L, Li W, Yi X, Sun L, Liang J, Cheng Z, Shi L, Shang Y, Yu W. SIRT7 is a histone desuccinylase that functionally links to chromatin compaction and genome stability. *Nat Commun.* 2016;7:12235.
19. Jiang Y, Liu J, Chen D, Yan L, Zheng W. Sirtuin inhibition: strategies, inhibitors, and therapeutic potential. *Trends Pharmacol Sci.* 2017;38:459–472.
20. Schiedel M, Robaa D, Rumpf T, Sippl W, Jung M. The current state of NAD^+-dependent histone deacetylases (sirtuins) as novel therapeutic targets. *Med Res Rev.* 2017.
21. Borra MT, Langer MR, Slama JT, Denu JM. Substrate specificity and kinetic mechanism of the Sir2 family of NAD^+-dependent histone/protein deacetylases. *Biochemistry.* 2004;43:9877–9887.
22. Smith BC, Denu JM. Sir2 protein deacetylases: evidence for chemical intermediates and functions of a conserved histidine. *Biochemistry.* 2006;45:272–282.
23. Hawse WF, Hoff KG, Fatkins DG, Daines A, Zubkova OV, Schramm VL, Zheng W, Wolberger C. Structural insights into intermediate steps in the Sir2 deacetylation reaction. *Structure.* 2008;16:1368–1377.

24. Jin L, Wei W, Jiang Y, Peng H, Cai J, Mao C, Dai H, Choy W, Bemis JE, Jirousek MR, Milne JC, Westphal CH, Perni RB. Crystal structures of human SIRT3 displaying substrate-induced conformational changes. *J Biol Chem.* 2009;284:24394–24405.
25. Smith BC, Denu JM. Mechanism-based inhibition of Sir2 deacetylases by thioacetyl-lysine peptide. *Biochemistry.* 2007;46:14478–14486.
26. Smith BC, Denu JM. Sir2 deacetylases exhibit nucleophilic participation of acetyl-lysine in NAD^+ cleavage. *J Am Chem Soc.* 2007;129:5802–5803.
27. Hu P, Wang S, Zhang Y. Highly dissociative and concerted mechanism for the nicotinamide cleavage reaction in Sir2Tm enzyme suggested by ab initio QM/MM molecular dynamics simulations. *J Am Chem Soc.* 2008;130:16721–16728.
28. Liang Z, Shi T, Ouyang S, Li H, Yu K, Zhu W, Luo C, Jiang H. Investigation of the catalytic mechanism of Sir2 enzyme with QM/MM approach: SN1 vs SN2? *J Phys Chem B.* 2010;114:11927–11933.
29. Cen Y, Sauve AA. Transition state of ADP-ribosylation of acetyllysine catalyzed by *Archaeoglobus fulgidus* Sir2 determined by kinetic isotope effects and computational approaches. *J Am Chem Soc.* 2010;132:12286–12298.
30. Zhou GC, Parikh SL, Tyler PC, Evans GB, Furneaux RH, Zubkova OV, Benjes PA, Schramm VL. Inhibitors of ADP-ribosylating bacterial toxins based on oxacarbenium ion character at their transition states. *J Am Chem Soc.* 2004;126:5690–5698.
31. Hirsch BM, Du Z, Li X, Sylvester JA, Wesdemiotis C, Wang Z, Zheng W. Potent sirtuin inhibition bestowed by L-2-amino-7-carboxamidoheptanoic acid (L-ACAH), a N^{ε}-acetyl-lysine analog. *Med Chem Commun.* 2011;2:291–299.
32. Zhou Y, Zhang H, He B, Du J, Lin H, Cerione RA, Hao Q. The bicyclic intermediate structure provides insights into the desuccinylation mechanism of human sirtuin 5 (SIRT5). *J Biol Chem.* 2012;287:28307–28314.
33. Wang Y, Fung YM, Zhang W, He B, Chung MW, Jin J, Hu J, Lin H, Hao Q. Deacylation mechanism by SIRT2 revealed in the 1′-SH-2′-O-myristoyl intermediate structure. *Cell Chem Biol.* 2017;24:339–345.
34. Cen Y, Falco JN, Xu P, Youn DY, Sauve AA. Mechanism-based affinity capture of sirtuins. *Org Biomol Chem.* 2011;9:987–993.
35. Seto E, Yoshida M. Erasers of histone acetylation: the histone deacetylase enzymes. *Cold Spring Harb Perspect Biol.* 2014;6:a018713.
36. Goodman M, Ro S, Peptidomimetics for drug design. 5th ed. Wolff ME, ed. *Burgers Medicinal Chemistry and Drug Discovery Vol 1 Principles and Practice.* John Wiley & Sons, Inc; 1995: 803–861.
37. Baba R, Hori Y, Mizukami S, Kikuchi K. Development of a fluorogenic probe with a transesterification switch for detection of histone deacetylase activity. *J Am Chem Soc.* 2012;134:14310–14313.
38. Wang Y, Chen Y, Wang H, Cheng Y, Zhao X. Specific turn-on fluorescent probe with aggregation-induced emission characteristics for SIRT1 modulator screening and living-cell imaging. *Anal Chem.* 2015;87:5046–5049.
39. Kawaguchi M, Ikegawa S, Ieda N, Nakagawa H. A fluorescent probe for imaging sirtuin activity in living cells, based on one-step cleavage of the dabcyl quencher. *Chembiochem.* 2016;17:1961–1967.
40. Xuan W, Yao A, Schultz PG. Genetically encoded fluorescent probe for detecting sirtuins in living cells. *J Am Chem Soc.* 2017;139:12350–12353.
41. Xie Y, Ge J, Lei H, Peng B, Zhang H, Wang D, Pan S, Chen G, Chen L, Wang Y, Hao Q, Yao SQ, Sun H. Fluorescent probes for single-step detection and proteomic profiling of histone deacetylases. *J Am Chem Soc.* 2016;138:15596–15604.
42. Kalesh KA, Tate EW. A succinyl lysine-based photo-cross-linking peptide probe for Sirtuin 5. *Org Biomol Chem.* 2014;12:4310–4313.

43. Yang T, Liu Z, Li XD. Developing diazirine-based chemical probes to identify histone modification "readers" and "erasers". *Chem Sci.* 2015;6:1011–1017.
44. Liu Z, Yang T, Li X, Peng T, Hang HC, Li XD. Integrative chemical biology approaches for identification and characterization of "erasers" for fatty-acid-acylated lysine residues within proteins. *Angew Chem Int Ed Engl.* 2015;54:1149–1152.
45. Feldman JL, Baeza J, Denu JM. Activation of the protein deacetylase SIRT6 by long-chain fatty acids and widespread deacylation by mammalian sirtuins. *J Biol Chem.* 2013;288:31350–31356.
46. Bao X, Wang Y, Li X, Li XM, Liu Z, Yang T, Wong CF, Zhang J, Hao Q, Li XD. Identification of "erasers" for lysine crotonylated histone marks using a chemical proteomics approach. *Elife.* 2014;3:e02999.
47. Jain AK, Barton MC. Bromodomain histone readers and cancer. *J Mol Biol.* 2017;429:2003–2010.
48. Zhang C, Zhai Z, Tang M, Cheng Z, Li T, Wang H, Zhu WG. Quantitative proteome-based systematic identification of SIRT7 substrates. *Proteomics.* 2017;17(13–14).
49. Smith BC, Settles B, Hallows WC, Craven MW, Denu JM. SIRT3 substrate specificity determined by peptide arrays and machine learning. *ACS Chem Biol.* 2011;6:146–157.
50. Bonkowski MS, Sinclair DA. Slowing ageing by design: the rise of NAD^+ and sirtuin-activating compounds. *Nat Rev Mol Cell Biol.* 2016;17:679–690.
51. Fatkins DG, Monnot AD, Zheng W. Nepsilon-thioacetyl-lysine: a multi-facet functional probe for enzymatic protein lysine Nepsilon-deacetylation. *Bioorg Med Chem Lett.* 2006;16:3651–3656.
52. Tong Z, Wang Y, Zhang X, Kim DD, Sadhukhan S, Hao Q, Lin H. SIRT7 is activated by DNA and deacetylates histone H3 in the chromatin context. *ACS Chem Biol.* 2016;11:742–747.
53. Costantini S, Sharma A, Raucci R, Costantini M, Autiero I, Colonna G. Genealogy of an ancient protein family: the sirtuins, a family of disordered members. *BMC Evol Biol.* 2013;13:60.

CHAPTER TWO

Targeting Sirtuins: Substrate Specificity and Inhibitor Design

Nima Rajabi, Iacopo Galleano, Andreas S. Madsen, Christian A. Olsen[1]
Center for Biopharmaceuticals, University of Copenhagen, Copenhagen, Denmark
[1] Corresponding author. E-mail address: cao@sund.ku.dk

Contents

Abstract

Lysine residues across the proteome are modified by posttranslational modifications (PTMs) that significantly enhance the structural and functional diversity of proteins. For lysine, the most abundant PTM is ε-*N*-acetyllysine (Kac), which plays numerous roles in regulation of important cellular functions, such as gene expression (epigenetic effects) and metabolism. A family of enzymes, namely histone deacetylases (HDACs), removes these PTMs. A subset of these enzymes, the sirtuins (SIRTs), represent class III HDAC and, unlike the rest of the family, these hydrolases are NAD^+-dependent. Although initially described as deacetylases, alternative deacylase functions for sirtuins have been reported, which expands the potential cellular roles of this class of enzymes.

Progress in Molecular Biology and Translational Science, Volume 154
ISSN 1877-1173
http://dx.doi.org/10.1016/bs.pmbts.2017.11.003

Currently, sirtuins are investigated as therapeutic targets for the treatment of diseases that span from cancers to neurodegenerative disorders. In the present book chapter, we review and discuss the current literature on novel ε-*N*-acyllysine PTMs, targeted by sirtuins, as well as mechanism-based sirtuin inhibitors inspired by their substrates.

1. INTRODUCTION

1.1 Structure and Classification of Sirtuins

Sirtuins are a class of enzymes characterized by sequence similarity to the yeast progenitor sirtuin *silent information regulator 2* (Sir2), which is named after its role in the regulation of mating type genes.[1,2] Seven sirtuins have been found in human cells (Fig. 1A) and they were classified as class III histone deacetylases (HDAC). Furthermore, phylogenetic analysis of sirtuins across multiple species grouped the enzymes in four main classes (I–IV) of eukaryotic enzymes, encompassing the human isotypes, as well as at least one more class containing enzymes found in archaea and bacteria (class U).[3] According to this classification, SIRT1–3 are part of class I, SIRT4 represents class II, SIRT5 belongs to class III, while SIRT6 and SIRT7 are members of class IV (Fig. 1B).[3,4] Sirtuins may also be discriminated by subcellular localization: SIRT1, 6, and 7 are nuclear; SIRT3–5 are mitochondrial; while SIRT2 is mainly cytosolic, but can shuttle to the nucleus during mitosis.[5,6] However, isoforms of SIRT1 and SIRT5 have also been identified in the cytoplasm.[7,8]

Sirtuins were initially thought to be mono-ADP-ribosyltransferases;[9,10] however, shortly after their discovery it was shown that sirtuins were able to deacetylate histone proteins in presence of NAD^+.[11–13] It was also proposed that histone proteins were not the only substrates of sirtuins,[13] mainly because most prokaryotes, which do not have histones, also express sirtuin enzymes. More insight into the role of sirtuins in human cells were provided after SIRT2 was crystallized and its structure elucidated:[14] a globular domain, containing the common Rossmann-fold domain for NAD^+ dependent enzymes, is connected by four extended β-strands to a smaller domain, where four cysteine residues coordinates a zinc atom (Fig. 1A and Fig. 1C). Sirtuins are not "zinc-dependent" like the other HDACs because the zinc atom does not take part in the catalytic mechanism of action. However, the zinc atom is important for the overall structure of sirtuins, which as a result affects the enzymes activity. The groove formed between the two domains is the site where the substrate and the cosubstrate bind (Fig. 1C).[15]

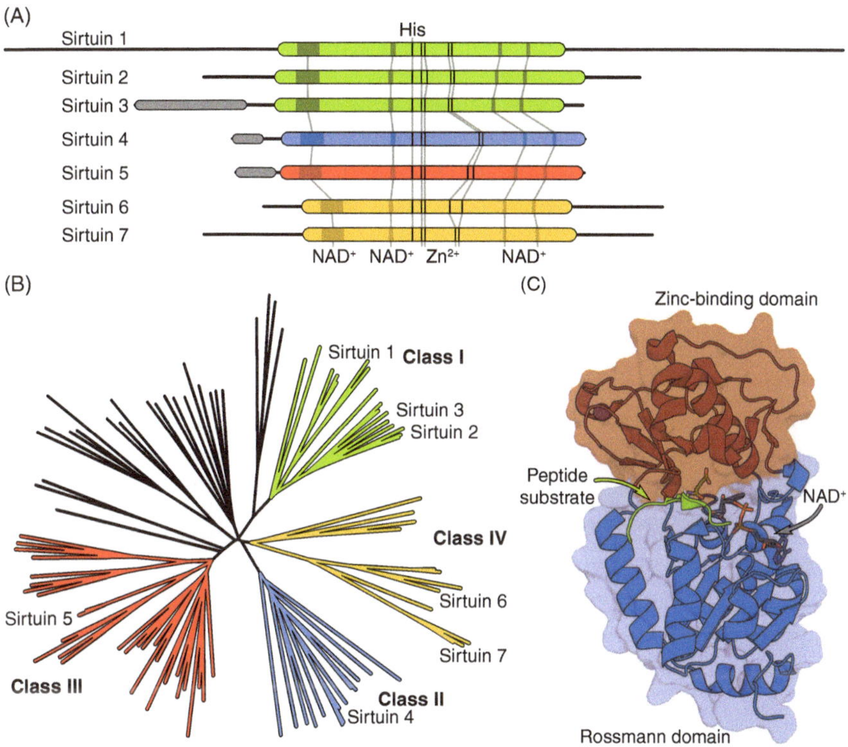

Fig. 1 Human sirtuins. (A) Schematic representation or human sirtuins, highlighting residues and regions of interest. Catalytic domains reflecting sirtuin classes are shown in *green* (class I), *blue* (class II), *red* (class III), and *yellow* (class IV); mitochondrial targeting sequence is shown in *gray*; NAD^+ binding regions are shown in a darker shade; the conserved catalytic histidine and zinc-coordinating cysteines are shown in *black*. (B) Phylogenetic tree of the NCBI Conserved Protein Domain Sir2-family (cd1407), highlighting classes (colors as aforementioned) and labeling the human sirtuins. (C) Crystal structure of a sirtuin 5-peptide substrate-NAD^+ complex (pdb 3riy), the zinc-binding and Rossmann domains are shown in *red* and *blue*, respectively, peptide substrate in *green*, and NAD^+ in *gray*.

1.2 Lysine Acylation

Even though sirtuins were initially viewed as deacetylases, it has been shown that sirtuins also target other ε-*N*-acyllysine posttranslational modifications (PTMs). Studies showing that sirtuins SIRT4, 5, and 7 exhibited weak deacetylase activity,[5,16] prompted researchers to search for alternative roles for these isotypes. The first identification of a nonacetyl deacylase activity for sirtuins was reported in 2011,[17] when SIRT5 was shown to catalyze the hydrolysis of ε-N-malonyllysine and ε-N-succinyllysine (Kmal and Ksuc, respectively, Fig. 2A).[15,18] This discovery motivated the scientific community to search for other acyl PTMs and to probe for new sirtuin deacylase activities. Later studies have identified other sirtuin targets, such as ε-*N*-glutaryllysine[19] (Kglu, Fig. 2A) and ε-*N*-myristoyllysine[20] (Kmyr, Fig. 2B), which are deacylated by SIRT5 and SIRT6, respectively, along with several other sirtuin substrates (Fig. 2B). Collectively, the posttranslational modifications in Fig. 2 represent the known ε-*N*-acyllysine landscape, and will be discussed in further detail later in the chapter in connection to the sirtuins catalyzing their hydrolysis.

Lysine acetylation is known to be catalyzed by lysine acetyltransferases (KATs), using acetyl coenzyme A (acetyl-CoA) as the acetyl donor. Moreover, KATs have been proven to catalyze the addition of other small acyl groups, such as propionyl, butyryl, and crotonyl (Fig. 2).[40,54] Therefore, the newly discovered ε-*N*-acyllysine PTMs were initially postulated to be formed by a similar mechanism.[40] Even though KATs have been found in both nucleus and cytoplasm, mitochondrial KATs are still elusive. Additionally, transferases that can catalyze acylation using other acyl-CoAs have not been identified.[55] For this reason, a nonenzymatic acylation mechanism has been proposed, where the lysine attacks the acyl-CoA thioester resulting in the acyl PTM. Even though nonselective acylation of a substrate can be envisioned without the participation of a transferase, it was recently shown that the acyl-CoAs show different reactivity toward distinct lysine residues.[56] Furthermore, nonenzymatic modification of lysine residues by succinyl-, glutaryl-, and 3-hydroxy-3-methylglutaryl-CoAs seems to occur via formation of the corresponding cyclic anhydride, which was shown to be more reactive than the CoA species.[57] A complementary nonenzymatic mechanism has also been proposed, where vicinal cysteine residues promote transfer of the acetyl moiety to lysine residues, via formation of a reactive acetyl cysteine thioester.[58] Regardless of the acylation mechanism, acyl-CoAs have been recognized as the source for many of these PTMs. Since the acyl-CoAs are involved in several metabolic pathways the energetic state of the cell could directly dictate different states of protein acylation.[23,59–61]

Along with specific acyl substrate recognition, it has been proposed that sirtuins can show selectivity based on substrate sequence. However, to date, there is no convincing evidence that the sirtuin isotypes have specific sequence recognition patterns. Some research groups have tried to elucidate flanking sequence preferences of the different sirtuin isotypes, but many of these independent studies have shown opposing or variable results.[62–67]

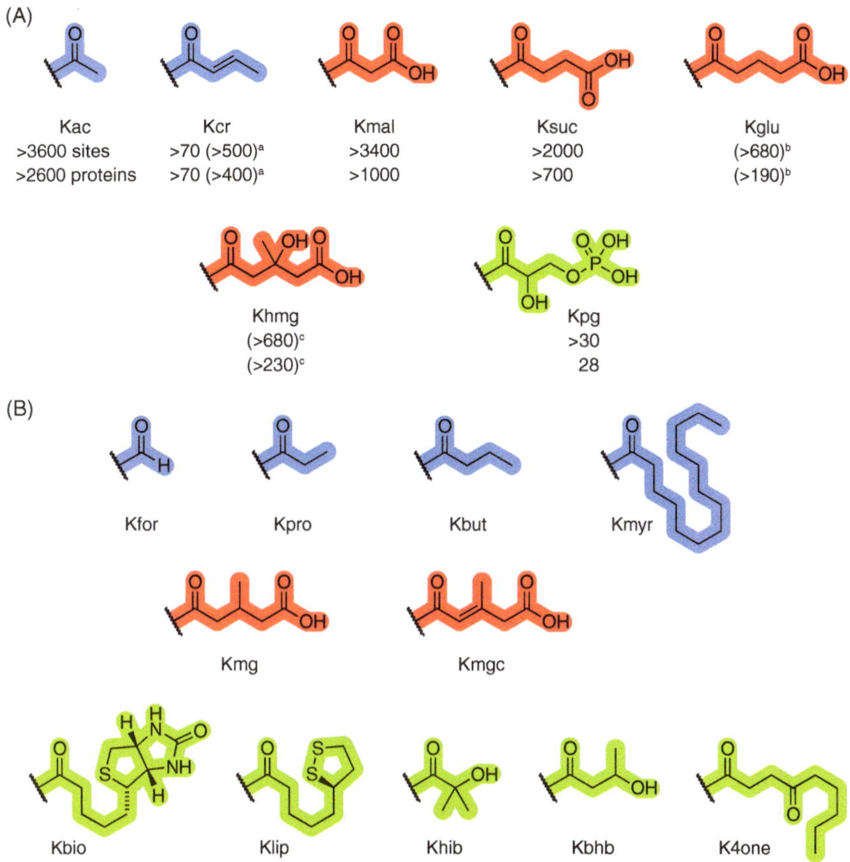

Fig. 2 ε-*N*-Acyllysine posttranslational modifications (PTMs) described in literature. (A) Proteome-wide analyses including aggregated numbers for modification sites and proteins. (B) Other acyl PTMs. [a]Numbers in parentheses are from samples grown in sodium crotonylate-enriched medium. [b]Numbers in parentheses are from *Sirt5*$^{-/-}$ mice liver. [c]Numbers in parentheses are from *HMGCL*$^{-/-}$(hydroxymethylglutaryl CoA lyase) mice liver. ε-*N*-Acetyllysine (Kac),[21–25] ε-*N*-crotonyllysine (Kcr),[26–30] ε-*N*-malonyllysine (Kmal),[15,17,31–33] ε-*N*-succinyllysine (Ksuc),[15,18,32–34] ε-*N*-glutaryllysine (Kglu),[19,33] ε-*N*-3-hydroxy-3-methylglutaryllysine (Khmg),[35] ε-*N*-3-phosphoglycerylysine (Kpg),[36] ε-*N*-formyllysine (Kfor),[37–39] ε-*N*-propionyllysine (Kpro),[40–42] ε-*N*-butyryllysine (Kbut),[40,42,43] ε-*N*-myristoyllysine (Kmyr),[20,44–46] ε-*N*-3-methylglutaryllysine (Kmg),[35] ε-*N*-3-methylglutaconoyllysine (Kmgc),[35] ε-*N*-biotinyllysine (Kbio),[47,48] ε-*N*-lipoyllysine (Klip),[47,49] ε-*N*-2-hydroxyisobutyryllysine (Khib),[50] ε-*N*-β-hydroxybutyryllysine (Kbhb),[51] ε-*N*-4-oxononanoyllysine (K4one)[52,53].

1.3 Sirtuin Activity Modulators

Since the discovery of the sirtuins, modulators of their enzymatic activity have been of great interest and numerous studies have led to the discovery of a vast array of such compounds. Sirtuin activity modulators have been used as tool compounds to further elucidate the biological roles of sirtuins in different tissues and diseases,[68,69] and may lead to innovative drugs for the treatment of a wide range of diseases.[70] Initially, sirtuin activators (Fig. 3A) received attention due to studies showing that overexpression or activation of Sir2 homologs led to an extended life span in diverse species, and loss of insulin resistance in diet-induced obese mice, fed on a high-fat diet (Fig. 3A).[71–75] In parallel, sirtuin inhibitors emerged as potential therapeutics, as sirtuins have been shown to be involved in cancer and neurodegenerative diseases.[76–79] So far, there are no approved sirtuin activity modulators on the market, but several compounds are currently undergoing clinical trials for the treatment of cancer, diabetes, or Huntington's disease (Fig. 3A and 3B).[80,81] Three major strategies have been utilized in the discovery of potent sirtuin inhibitors; chemical library screenings[76,82,83] (Fig. 3B), structure-based virtual screening[84–87] (Fig. 3B), and the design of mechanism-based (peptidomimetic) inhibitors (Fig. 3C). Although there has been great progress in the field, there is still a need for more potent and more selective inhibitors. In this chapter, we will focus on mechanism-based inhibitors, referring readers to several excellent reviews that cover the other strategies.[70,88–91] Herein, we will illuminate the rationale that researchers have applied in their design of mechanism-based inhibitors and discuss how these compounds inhibit sirtuin activity. The typical structure of mechanism-based sirtuin inhibitors is based on a lysine motif, which carries an amide isostere, an acyl chain analog, and has a peptidic-, pseudo-peptidic, or macrocylic scaffold (Fig. 3C). These different structural elements all contribute, to some degree, to the selectivity, binding affinity, and potency of the inhibitors against the different sirtuin isotypes.

1.4 Mechanism

Understanding the sirtuin catalyzed deacylation reaction is crucial in the development of potent inhibitors and much of the mechanism has actually been revealed by the use of inhibitors.[95–99] Since the publication of the first sirtuin crystal structure, efforts have been undertaken to elucidate the molecular mechanism of sirtuin deacylation. Initially, it was not clear which substrate binds first, but studies have suggested that binding of peptide substrate could induce a conformational change, which then allows binding

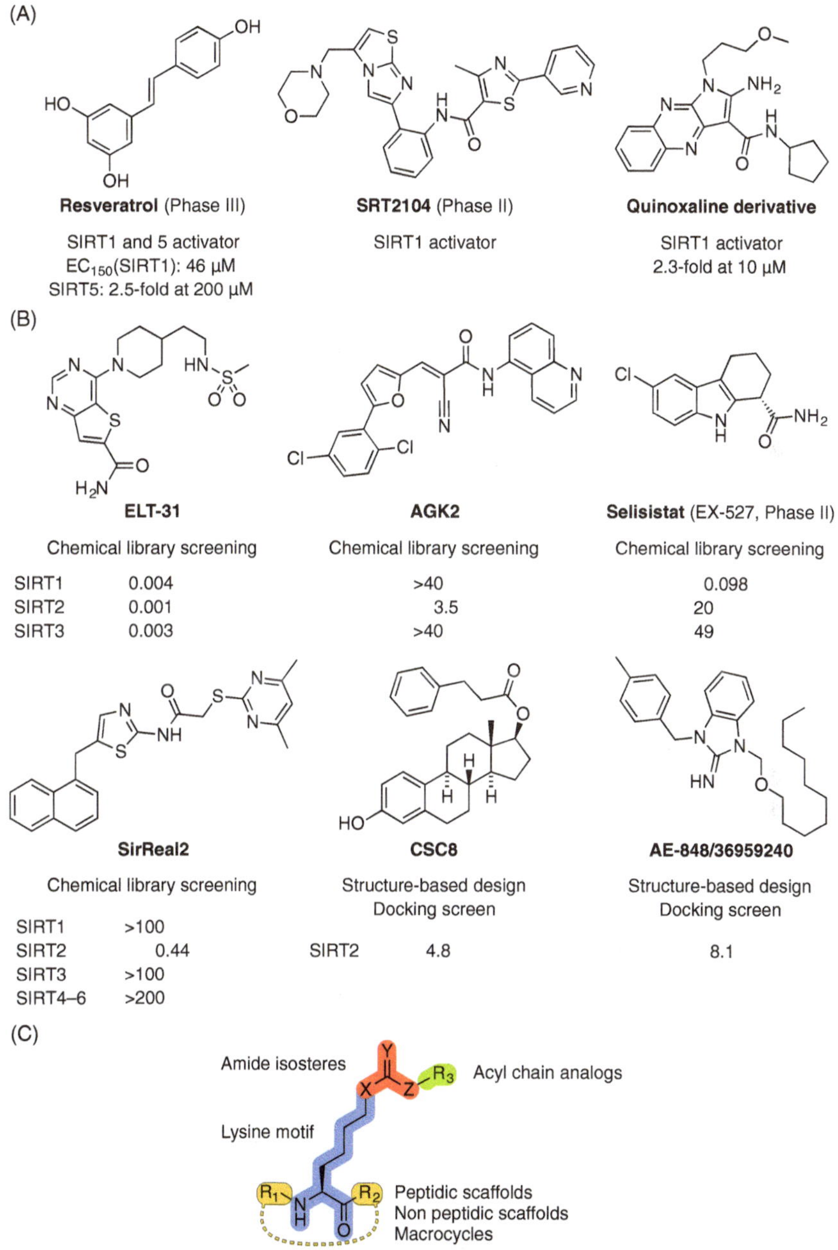

Fig. 3 Activators, small molecule inhibitors, and mechanism-based inhibitors. (A) Selected sirtuin activators, their target, and activation potential. (B) Selected nonpeptidomimetic inhibitors, means of identification, and IC_{50} values (μM). (C) General pharmacophore for mechanism-based inhibitors. **Reveratrol**,[71] **SRT2104**,[92] **Quinoxaline**,[93] **ELT-31**,[82] **AGK2**,[76] **Selisistat** (EX-527),[83] **SirReal2**,[94] **CSC8**,[84] **AE-848**[86].

of the cosubstrate.[14,100] Furthermore, cocrystallization studies of the bacterial Sir2Tm and the archeal sirtuin Sir2Af2, with acetylated peptides showed that the substrate:enzyme complex had a more closed conformation compared to the free enzyme.[62,101] Eventually, studies on SIRT2, Sir2Af2, and the yeast sirtuin HST2 showed that NAD^+ binds in a productive way, only upon initial binding of the ε-*N*-acyllysine substrate.[101,102]

After binding of the two substrates, the sirtuin catalyzed deacylation reaction is initiated by attack of the carbonyl oxygen on the anomeric carbon of the nicotinamide ribose ring (C1′), forming a 1′-O-alkylimidate intermediate (**I**, Fig. 4A). This has been confirmed in various studies, and the intermediate was even captured in a crystal structure.[103,104] Using thioamide-based sirtuin inhibitors, the resulting 1′-S-alkylimidate is longer-lived and has also been captured in crystal structures (**I**, Fig. 4B).[99,105,106] A concerted mechanism has been proposed, where the carbonyl oxygen is involved in direct attack on the anomeric carbon of the ribose ring.[97] Others have suggested a stepwise reaction, involving the formation of a positively charged oxocarbenium intermediate as a results of nicotinamide dissociation. This has been supported by crystal structures showing that the distance between the oxygen of the acetyllysine and the C1′ of the ribose was not compatible with a direct nucleophilic attack.[107,108] Another study showed that the active site can accommodate various NAD^+ conformations, and that the C1′ of the ribose is likely to move closer to the carbonyl oxygen during catalysis.[109] Whether the 1′-O-alkylimidate is formed through a concerted or stepwise mechanism, or a mechanism with characteristics of both, is still under debate.[110]

The forward reaction is catalyzed by a conserved histidine residue (Fig. 1A), which deprotonates the 2′-OH of the 1′-O-alkylimidate intermediate, which then attacks the imidate carbon to form a bicyclic intermediate (**II**, Fig. 4A). Sirtuin mutants lacking this conserved histidine show significant reduction in enzymatic activity, affirming its importance in catalysis.[104,109,111] Using a thioamide-based inhibitor, the resulting bicyclic intermediate has also been trapped in a crystal structure (**II**, Fig. 4B).[98] Although conclusive experimental evidence is lacking, it has generally been thought that water can attack the bicyclic intermediate directly, resulting in the release of deacylated peptide. However, in a recent study, Lin and coworkers captured a new intermediate, a 1′-SH-2′-O-alkylimidate, in a crystal structure of SIRT2 with an ε-*N*-thiomyristoyl inhibitor (**III**, Fig. 4B).[99] This suggests that the bicyclic intermediate of the sirtuin-catalyzed deacylation reaction most likely decomposes, resulting in a 1′-OH-2′-O-alkylimidate

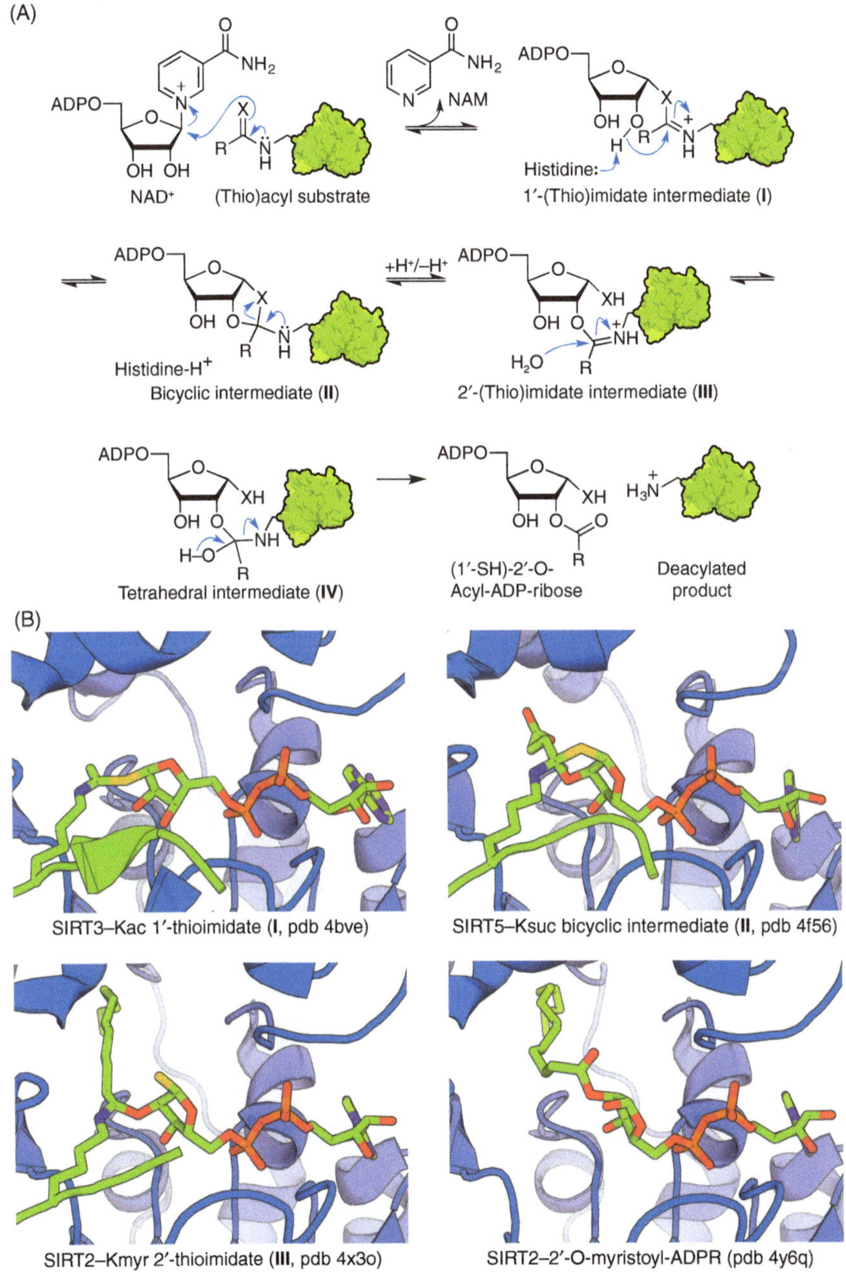

Fig. 4 Putative mechanism and crystal structures. (A) Putative mechanism for sirtuin catalyzed deacylation of an ε-*N*-(thio)acylated protein (shown in *green*, X=O or S). (B) Cocrystal structures of sirtuins with the three types of intermediates reported and the acylated ADP-ribose product.

intermediate. Subsequent hydrolysis of the 1′-OH-2′-O-alkylimidate and decomposition would result in the release of the deacylated lysine and the 2′-O-acyl adenosine diphosphate ribose (2′-OAADPR) (Fig. 4A). This proposed mechanism is in agreement with experimental data from an earlier study by Smith and Denu, using $H_2{}^{18}O$ to probe the hydrolysis of the bicyclic intermediate.[104] Although they proposed the initially suggested mechanism, where water attacks the bicyclic intermediate, they showed that in lysine deacetylation by HST2, ^{18}O was exclusively incorporated as the carbonyl oxygen of the 2′-OAADPR. This would also be seen if the bicyclic intermediate would initially decompose to the 1′-OH-2′-O-alkylimidate intermediate, followed by attack of water on the imidate carbon to form the tetrahedral intermediate (**IV**, Fig. 4A). It is therefore plausible to assume that the natural substrate is processed through the same intermediates as the ε-*N*-thiomyristoyl inhibitor reported by Lin and coworkers.[99]

Clearly, considerable efforts have been made to elucidate the sirtuin catalyzed deacylation reaction, and the progress over the years has now led to a proposed mechanism which has been substantiated by the collective results of numerous studies. This has been a crucial part in the general understanding of sirtuins and has facilitated the design of mechanism-based inhibitors.

2. MECHANISM-BASED SIRTUIN INHIBITION

As the carbonyl oxygen of the ε-amide is directly involved in the initial step in the sirtuin deacylation reaction, inhibitors have been designed with various amide isosteres (Fig. 5). These can affect this step in the reaction in different ways, which we will discuss in more detail further.

2.1 Thioamide- and Thiourea-Based Inhibitors

Numerous sirtuin inhibitors have been based on the thioamide moiety since the mid-2000s[112] (Fig. 5), which has resulted in some of the most potent sirtuin inhibitors to date. In 2007, Smith and Denu showed that a

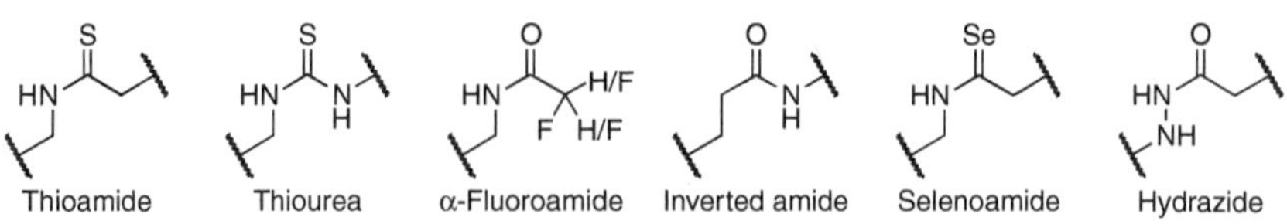

Fig. 5 ε-*N*-Amide isosteres used in mechanism-based inhibitors.

ε-*N*-thioacetyllysine (Ktac) peptide inhibitor binds with higher affinity to HST2, compared to its corresponding ε-*N*-acetyllysine (Kac) peptide.[95,96] Moreover, they showed that the Ktac inhibitor had an 80-fold decrease in overall turnover compared to the Kac substrate. As the rate of nicotinamide formation was very similar between the two species, they concluded that the formation of the bicyclic intermediate had to be slower for the Ktac inhibitor. Other groups have validated the formation of this stalled 1′-S-alkylimidate intermediate over the last decade (**I**, Fig. 4B).[99,105,106] This intermediate has been suggested to be stalled because the longer bond length of S—C compared to O—C results in an increased distance of the 2′-hydroxyl group to the carbonyl carbon and possibly a nonoptimal angle for attack on the imidate carbon.[95] Thus, formation of the bicyclic intermediate is slower, but it is possible that the subsequent step, that is, formation of the 1′-SH-2′-O-alkylimidate is slower as well (**III**, Fig. 4B). Ring opening of the bicyclic intermediate is most likely a reversible reaction, and it is plausible that the more nucleophilic thiol would reform the bicyclic ring more readily than the hydroxyl. Furthermore, as the 2′-alkylimidate and the bicyclic intermediate have only been reported in one crystal structure each, it could be reasoned that the overall equilibrium of both reactions lies toward the 1′-S-alkylimidate (**I**, Fig. 4B).

While the thioamide is one of the most successful amide isosteres in mechanism-based inhibitor design, it has been argued that possible cytotoxicity issues, especially hepatotoxicity, might hinder the therapeutic potential of such inhibitors.[113] Consequently, a close analog to the thioamide, the thiourea moiety (Fig. 5), has been introduced as an alternative to potentially circumvent this issue.[113–115] Thiourea-based inhibitors have also been shown to exhibit potent sirtuin inhibition.[114,116,117] They have been proposed to be processed in a manner similar to the thioamides, to produce a stalled 1′-S-alkylisothiourea intermediate after the initial step in the catalytic cycle.[113] This intermediate is thought to be longer-lived than the 1′-S-alkylimidate generated from a thioamide.[113] In the HST2-catalyzed reaction, Smith and Denu also showed that a homocitrulline-containing peptide could form a stalled 1′-O-alkylisourea intermediate, which did not proceed to form the free lysine ε-amine. Instead, this intermediate was hydrolyzed and reverted back to the initial inhibitor, only forming nicotinamide and adenosine diphosphate ribose as products.[96] This could be possible for thiourea-based inhibitors as well, since there have been no reports showing sirtuins-catalyzed deacylation of thiourea compounds, and no cocrystal structures of thiourea compounds bound to sirtuins have yet been published. If this is true, this will prove to be advantageous in terms of efficacy of the inhibitors, as reformation of the inhibitor would allow it to inhibit the sirtuin deacylation reaction repeatedly. Additionally, thiourea-based sirtuin inhibitors have been shown to possess antioxidant activity against superoxide radical anions, which could provide additional therapeutic effects.[115]

The potent sirtuin inhibition exhibited by thioamides and thioureas is generally unchallenged by any other amide isosteres, with only a few exceptions. Consequently, most of the recently reported mechanism-based inhibitors have been based on these isosteres.

2.2 α-Fluoro Amide Inhibitors

Another amide isostere that has been used in the design of sirtuin inhibitors is α-fluoro amides (Fig. 5). ε-*N*-fluoroacetyl groups have been shown to bind to sirtuins with equal or higher affinity compared to ε-*N*-acetyllysine. Supposedly, this is because of the increased hydrophobicity and size of the fluoroacetyl group, which fills the binding pocket to a higher extent compared to acetyl.[97] Additionally, ε-*N*-fluoroacetylated peptides have been shown to have a slower rate of nicotinamide formation, presumably due to the decreased nucleophilicity of the carbonyl oxygen. This is linked to the electron-withdrawing effect of the α-fluorine atoms, evident by mono-, di-, and trifluoroacetyl derivatives, which show decreasing rates of nicotinamide formation with increasing number of α-fluorine atoms.[96,97] The decreased nucleophilicity of the carbonyl oxygen is, however, also accompanied by an increased electrophilicity of the carbonyl carbon. It could therefore be argued that while the initial attack on the ribose ring is slower, the formation of the bicyclic intermediate would occur faster. The increased binding affinity and slow turnover rate of these compounds suggests that this amide isostere have the potential for generating highly potent inhibitors. Although initial studies showed promise for fluoroacyl groups as suitable amide isosteres for mechanism-based inhibition, there are only few examples of such sirtuin inhibitors.[95,118]

2.3 L-α-Aminosuberic Acid-Based (Inverted Amide) Inhibitors

In an attempt to retain potency, sirtuin selectivity, and simultaneously avoid the potential toxicity accompanied with sulfur-containing carbonyl isosteres, Hirsch *et al.* have introduced inhibitors based on L-α-aminosuberic acid (L-Asu), resulting in an “inverted amide” (Fig. 5).[119] Through MALDI experiments, they showed that this analog could be processed similarly to the natural amide substrates, but instead leads to formation of a stalled intermediate. Although the structure of this intermediate could not be definitively determined, they hypothesized that the species observed was a bicyclic intermediate similar to that produced from the natural substrate.[119]

In 2015, He *et al.* published a series of L-Asu-based inhibitors, including amide-, oxy-, and thioester analogs.[120] Not surprisingly, due to the lower nucleophilicity of the carbonyl oxygen of the ester and thioester, the amide analogs exhibited more potent inhibition than the former. While the reported L-Asu inhibitors generally show significantly lower potencies compared to their thiourea or thioamide counterparts, further studies of L-Asu-based inhibitors should be considered, because it could potentially generate safer mechanism-based sirtuin inhibitors.

2.4 Miscellaneous and Potential Amide Isosteres

2.4.1 Selenoamides

Although only a few examples have been reported, selenoamide-based compounds (Fig. 5) have shown promise in sirtuin inhibitor design. Two studies have demonstrated that selenoamide-based inhibitors possess similar or slightly increased potency compared to their Ktac counterparts.[121,122] This might not be surprising since sulfur and selenium exhibit very similar physical and chemical properties, for example, atomic radius, electronegativity, and bond lengths.[123] Huhtiniemi *et al.* proposed that selenoamide inhibitors could potentially form a stalled 1′-Se-alkylimidate; however, no experimental data have confirmed this hypothesis yet.[121] Although selenoamides have been shown to exhibit effective sirtuin inhibition, the potential toxicity of these compounds may limit the value of selenoamide sirtuin inhibitors in biomedical applications.

2.4.2 Hydrazides

In 2012, Dancy *et al.* reported a peptide containing a hydrazide analog (Fig. 5) of Kac. When deacetylated by Sir2Tm or SIRT1, this peptide showed 8- and 18-fold slower turnover, respectively, compared to its Kac counterpart.[124] Although, the hydrazide-containing peptide showed no inhibition against Sir2Tm or SIRT1, it was shown that the binding affinity to Sir2Tm was sevenfold higher than the Kac equivalent. Moreover, subsequent to deacetylation of the hydrazide, they could isolate a small amount of a new stable hydrazide-ADP-ribose adduct. NMR analysis showed that this adduct contained a bond between the δ-nitrogen of the hydrazide and the

C1 of the ribose ring.[124] Further studies into hydrazide sirtuin inhibitors may be warranted as the high affinity, slow conversion rate, and possibility of this analog to form the stable adduct suggest that this isostere could have potential in sirtuin inhibitor design.

3. ε-*N*-ACYLATED LYSINES AND INHIBITORS

The sirtuin isotypes show distinct preference for ε-*N*-acyllysine PTM substrates. SIRT1–3 and SIRT6 have been shown to hydrolyze aliphatic acyl groups, with chain lengths ranging from 1 to 16 carbon atoms.[125–127] Carboxyacyl groups are mainly deacylated by SIRT4 and SIRT5,[15,19,35,128] whereas SIRT7 has not yet been shown to have any specificity and exhibits deacylation of several acyl PTMs.[129–131]

As the substrate selectivity is mainly determined by the acyl group, mechanism-based inhibitors are often designed to resemble these PTMs. Regardless of the amide isostere chosen in the development of mechanism-based inhibitors, isotype selectivity has so far mainly been achieved by exploiting the preference of sirtuins for these different acyl groups.

As lysine acylation is widespread in the cell and can lead to activation or inactivation of an enzymatic pathway, it is not surprising that sirtuin inhibition can result in either beneficial or detrimental effects, depending on disease and cell line. We would like to refer readers to several comprehensive reviews that cover this ambiguity of the therapeutic potential of sirtuins.[68,70,88,132] In this section, we will discuss selected examples where sirtuin activity against the ε-*N*-acyllysine PTMs is linked to diseases and highlight mechanism-based inhibitors that have been designed to target the individual sirtuin deacetylases.

3.1 ε-*N*-Acetyllysine

Acetylation is by far the most studied acyl group found on lysine residues. Historically, acetyl was the first acyl group found to decorate histones, when Allfrey *et al.* in the 1960s proposed that lysine acetylation affects RNA synthesis at different loci.[133] Acetylation of the ε-amino group of lysines neutralizes the positive charge that the ε-nitrogen ($pK_a \sim 10$) bears at physiological pH (pH ~ 7.4). In histones, this alteration of electrostatic properties results in a weaker interaction with DNA. However, Allfrey's results indicated that this effect could not fully account for the increased RNA expression. It took more than 30 years to gain more insight into how histones and chromatin are affected by acetylation. It has been shown, that proper folding of nucleosomes depend on the level of histone acetylation[134] and, from later experiments, that acetylation of a specific residue (lysine 16 of histone 4, H4K16) was enough to prevent proper folding.[135] The first proteomic investigation of acetylated lysines was conducted in 2006, where it was shown that ε-*N*-acetyllysines were ubiquitous in the cell.[22] These results were later confirmed and expanded in other proteomic studies.[21,25,136,137] As lysine acetylation is wide-spread, it is not surprising that this PTM affects and regulates many of the cellular processes. This is in accordance with the fact that acetyl-CoA, the metabolite recognized as the acetyl donor, is produced and consumed in many processes within the cell. Furthermore, acetyl-CoA levels vary depending on the energetic state of the cell. In mammalian cells, acetylation has been demonstrated to regulate metabolic enzymes.[23] Among these enzymes, malate dehydrogenase (MDH), the enzyme involved in the conversion of malate into oxaloacetate in the TCA cycle, and enoyl-CoA hydratase/3-hydroxyacyl-CoA (EHHADH), an enzyme involved in β-oxidation of fatty acids, have been found to be activated by acetylation. However, acetylation decreases the activity of argininosuccinate lyase (ASL), an important enzyme found in the urea cycle and involved in the synthesis of arginine.[23] These examples underline the dichotomous effect that acetylation can have on enzyme activity, which makes the effect of acetylation of a protein hard to predict. Among sirtuins, SIRT1–3 have been shown to be robust deacetylases, both in vitro and in vivo. Remarkably, SIRT1–3 show higher activity as long-chain deacylases in vitro.[125–127] However, their physiological role as long-chain deacylases has not been demonstrated so far.

3.1.1 SIRT1 as Deacetylase

SIRT1 is by far the most studied sirtuin. The fact that upregulation of the enzyme or its homologs leads to increased life-span in several models[71,72,74,75,138,139] has prompted researchers to investigate SIRT1 activation as a pharmacological approach. In neurodegenerative disorders, such as Parkinson's and Alzheimer's disease, SIRT1 has been shown to have a general protective effect.[140] Interestingly, however, the selective SIRT1 inhibitor Selisistat (Fig. 3B) is currently under clinical trials for the treatment of Huntington's disease.[80] As the transcription factor p53 was one of the first SIRT1 substrates described,[141] it is also one of the most studied SIRT1 targets. Loss-of-function mutations of p53 have been identified in over 50% of all human tumors.[142] Acetylation of p53 in chronic myelogenous leukemia cells caused by SIRT1 inhibition resulted in increased apoptosis, potentially mediated by the proapoptotic protein Bax. Apoptosis was further increased when the tyrosine kinase inhibitor imatinib was coapplied.[143] Downregulation of SIRT1 could be beneficial in other cancer cell lines as well.[132,144] Inhibition of SIRT1 enhances degradation of N-Myc, a transcription factor involved in the malignant transformation of neuroblast cells,[145] and leads to cell growth arrest in neuroblastoma.[146] SIRT1 activity has also been shown to affect the tumor suppressor retinoblastoma protein (Rb), which is posttranslationally modified by acetylation and phosphorylation.[147] Acetylation of Rb, which can be reverted by SIRT1, was found to be correlated with density-dependent growth inhibition of cells in culture.[148] Inhibition of SIRT1 could therefore represent a tool for tuning cell cycle growth/arrest in diseased cells.

3.1.2 SIRT2 as Deacetylase

In general, SIRT2 is recognized as a tumor suppressor.[132] However, there are examples where SIRT2 inhibition have been shown to have potential in targeting various cancers and neurodegenerative disorders.[149] Apoptosis in HeLa cells, a cervical cancer cell line, was induced upon inhibition of SIRT2, which in turn determined the downregulation of p300 and MDM2, a negative regulator of p53, eventually leading to accumulation of p53.[150] Intriguingly, direct knockdown of p300 led to a lower increase of p53 expression and lower degree of cell death than knockdown of SIRT2. Therefore, it is plausible that p53 accumulation and eventually apoptosis are effects derived from multiple pathways. Inhibition of SIRT2 has been shown to prevent the translocation of the transcription factor SREBP-2 into the nucleus, where it binds to DNA and eventually induces the synthesis of cholesterol.[151] Decreased accumulation of cholesterol led to a neuroprotective effect and improved cell viability in models of Huntington's disease (HD).[152] SIRT2 inhibition also reduced the number of mutant huntingtin inclusions, a protein involved in the etiology of HD. SIRT2 has also been linked to neurodegenerative disorders because it deacetylates the nuclear factor-kappa B (NF-κB).[149] Similarly, SIRT2 inhibition led to reduction in number and increase in size of α-synuclein aggregates, which in turn resulted in a cytoprotective effect in Parkinson's disease models.[76] Although deacetylation of α-tubulin has been suggested as a plausible effector, it is possible that a combination of SIRT2-dependent pathways contribute to the overall protective effects observed.

3.1.3 SIRT3 as Deacetylase

As SIRT3 is the only robust deacetylase in mitochondria,[153] its patophysiological role in cells has been of great interest. SIRT3 knockout (KO) mice, unlike SIRT1 KO animals, survive with no apparent malformations, which may suggest that SIRT3 is more involved in the protection of cells than in the development of the organism.[153] Downregulation of SIRT3, which is overexpressed in oral squamous cell carcinomas, resulted in decreased cell growth and increased sensitivity to both cisplatin and radiation.[154] Mechanistically, SIRT3 has been shown to have a protective role in stress-mediated cell death by deacetylation of Ku70, a protein which sequesters the proapoptotic protein Bax in cytoplasm.[155] Increased acetylation levels of Ku70 result in a lower degree of association with Bax,[156] suggesting that inhibition of SIRT3 could lead to hyperacetylation of Ku70 and eventually induce apoptosis. SIRT3 activity has in general been shown to be important for reducing apoptosis in cells under oxidative and genotoxic stress conditions.[157] It is plausible that cancer cells promote their survival by exploiting this mechanism. Accumulation of reactive oxygen species (ROS) is known to be toxic to cells and inhibition of SIRT3 could therefore have potential in treatment of cancer by modulating this mechanism.[158,159] As a result, the development of SIRT3-selective inhibitors would be of great value both as pharmacological probes and potential drug candidates.[159]

3.1.4 SIRT6 as Deacetylase

SIRT6 was initially reported to be a mono-ADP-ribosylase and deacetylase[160,161] involved in several mechanisms, such as telomer maintenance, DNA repair, and metabolic homeostasis.[161–164] Similar to other sirtuins, SIRT6 has been proposed to act as beneficial modulator and tumor suppressor in cells.[132,165] However, SIRT6 has been found to be overexpressed in several hematopoietic cancer cell lines, multiple myeloma, and solid tumors, thereby functioning as an oncogene. In nonsmall cell lung cancer cells (NSCLC), SIRT6 is potentially an oncogene. Enhanced degradation of SIRT6 upon activation of cAMP signaling resulted in an increased sensitivity to γ-ray induced apoptosis.[166] SIRT6 overexpression was also shown to be detrimental in human prostate cancer cells,[167] in squamous cell carcinomas[168] and in hepatocellular carcinomas,[169] where SIRT6 upregulation in general translates into increased cell proliferation and cell survival. Hence, in these tumorigenic cell lines, SIRT6 inhibition could offer a novel therapeutic target.

3.1.5 SIRT7 as Deacetylase

Although not robustly demonstrated in vitro, SIRT7 has also been identified as a deacetylase. SIRT7 deacetylates both histone and nonhistone proteins, such as PAF53, a subunit of RNA polymerase I (Pol I),[170,171] and U3-55k, a protein involved in pre-rRNA processing.[172] SIRT7 deacetylation of H3K9 and H3K18 could be achieved in vitro in presence of double stranded DNA, which has been proposed to interact with N- and C-termini of SIRT7.[130] Due to its role in RNA processing, it is not surprising that overexpression of SIRT7 can be found in cancers, where the translation machinery is highly altered. Overexpression of SIRT7 has been reported in some cancer cell lines, such as colorectal cancer[173] and breast cancer.[79] However, on a molecular level, only few substrates of SIRT7 have been discovered so far.[174] One of these targets is H3K18, whose deacetylation mediated by SIRT7 is associated with aggressive tumors.[175] Despite its important role in cells, identification of SIRT7 substrates is still lagging behind compared to the other sirtuins, mainly due to the low activity of the recombinant enzyme in vitro.

3.1.6 Inhibiting Deacetylation

As SIRT1–3 are the most robust deacetylases inhibition of these three isotypes has gained considerable attention. Given the fact that SIRT1–3 show similar ε-*N*-acyllysine preferences, designing selective mechanism-based inhibitors against these isotypes is a difficult task to accomplish. Numerous mechanism-based inhibitors containing Kac-mimicking analogs such as -Ktac or ε-*N*-trifluoroacetyllysine (Ktfa), have been designed to target SIRT1–3.[95,112,113,119,121,176–180] Many of these have been shown to reduce acetylation of, for example, p53, and some have also been shown to decrease proliferation of various cancer cells, highlighting their potential as therapeutics.[105,113,119,180,181] While these inhibitors often show high potency, the compounds are usually only moderately selective toward one of the SIRT1–3 isotypes.

To alter the potency and selectivity profiles of acetyl-analog inhibitors, various scaffolds have been investigated. Initially, sirtuin inhibitors were designed using peptidic scaffolds up to 20 amino acids in length, with sequences derived from known sirtuin targets, such as p53, H3K9, or α-tubulin (Fig. 6A).[95,112,176,178] The basis of using a scaffold that mimics the natural substrate as much as possible was to ensure binding affinity and thus gain potency. However, extensive studies from several research groups in the field have shown that this is not always the case.[105,121,180,182] Furthermore, as mentioned earlier, the sirtuins do not appear to have any specific sequence recognition pattern but rather binds via interactions with substrate backbone amides. Consequently, predicting binding affinities and selectivity profiles when designing peptidic scaffolds is very difficult. As seen

(A)

	SIRT1	SIRT2	SIRT3	SIRT5–7
H-**KRLPKTRSGKtacVMRRLLRKII**-OH (**1**)	0.9	4.3	4.5	ND
H-**KSTGGKtacAPRKQ**-OH (**2**)	2.0	5.6	2.3	ND
H-**HKKtacAM**-OH (**3**)	0.18	3.8	ND	ND
Ac-**AKtacA**-OH (**4**)	0.24	9.8	ND	ND
Ac-**NapKtacKac**-NH_2 (**5**)	12.6	12	1.29	ND
Ac-**RIKKtfaRY**-NH_2 (**6**)	0.28	0.031	1.0	ND

(B)

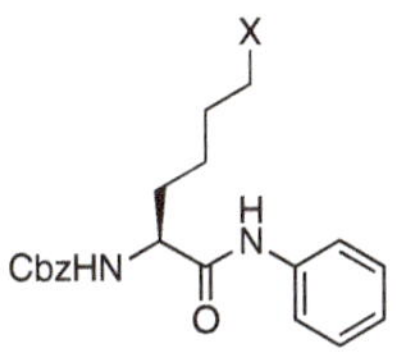

X =

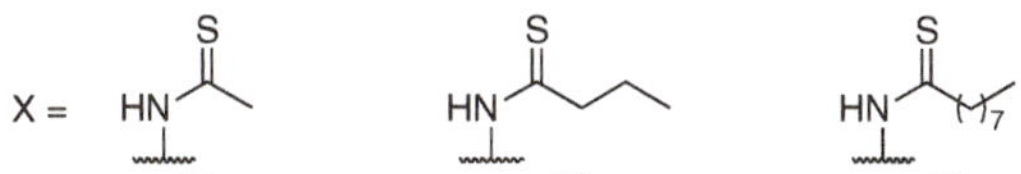

	(**7**)	(**8**)	(**9**)
SIRT1	2.7 / 10.4	3.8	1.2
SIRT2	23 / 5.5	0.43	0.13
SIRT3	>100 / 27.8	71.5	>200
SIRT5	ND / >200	>200	>200
SIRT6	ND / >200	>200	>200
SIRT7	ND / >200	>200	>200

X =

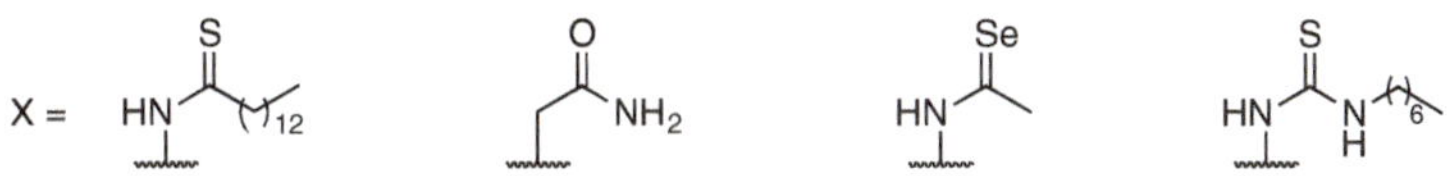

	(**10**)	(**11**)	(**12**)	(**13**)
SIRT1	98	>1000	9.9	64
SIRT2	0.028	ND	15	0.2
SIRT3	>200	ND	ND	>256
SIRT5	>200	ND	ND	>256
SIRT6	>200	ND	285	ND
SIRT7	>200	ND	ND	ND

(C) (D)

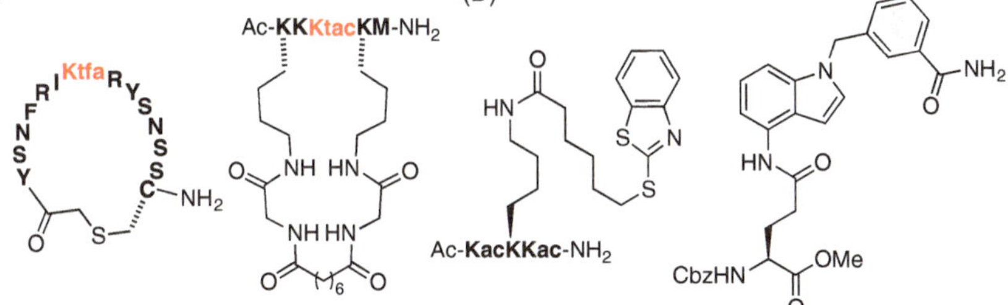

	(**14**)	(**15**)	(**16**)	(**17**)
SIRT1	0.047	0.0023	12.4	67
SIRT2	0.0032	0.144	>200	33
SIRT3	0.480	0.0429	>1000	3.6
SIRT5–7	ND	ND	ND	ND

Fig. 6 Structures and IC_{50} values (μM) of sirtuin inhibitors containing ε-*N*-acyllysine-mimicking motifs. Inhibition of desuccinylation (SIRT5), deacetylation (SIRT1–3), and deacetylation/demyristoylation (SIRT6) is shown. (A) Linear peptide inhibitors. (B) Lysine-derived inhibitors with various amide isosteres and acyl mimicks. (C) Cyclic peptide inhibitors. (D) Inhibitors targeting binding sites for both substrate and NAD^+; ε-*N*-thioacetyllysine (Ktac), ε-*N*-trifluoroacetyllysine (Ktfa). **1**,[176] **2**,[95] **3**,[178] **4**,[121] **5**,[184] **6**,[118] **7**,[105,182] **8-10**,[182] **11**,[119] **12**,[121,122] **13**,[114] **14**,[118] **15**,[185] **16**,[186] **17**[187].

in Fig. 6A, different lengths of peptides and various flanking residues often result in small changes in potencies and selectivity. Additionally, long peptides are known to possess poor cell permeability and metabolic stability,[183] and the focus has shifted to shorter peptide scaffolds and/or small pseudopeptidic scaffolds. For example, compounds with small pseudopeptidic scaffolds, such as thioamide **7** (Fig. 6B) have been shown to exhibit potencies equivalent to peptide scaffolds containing the same amide isostere (Fig. 6A).[105,121,180,182]

Another common strategy to increase potencies and alter selectivity profiles of SIRT1–3 inhibitors has been the use of different amide isosteres and acyl chain analogs. As seen in Fig. 6B, the choice of amide isostere can drastically affect potency and selectivity of the inhibitors. Moreover, the length of the acyl chain analog can have similar effects, as binding affinity of longer chain acyl groups differ between SIRT1–3. In 2016, Jing *et al.* reported compound **10** (Fig. 6B), a highly selective SIRT2 inhibitor.[182] This inhibitor was shown to have a broad anticancer effect, inhibiting growth of 36 different cell lines by indirectly inducing c-Myc degradation.

Macrocyclic scaffolds have also been used in sirtuin inhibitors, some of which have shown potent inhibition. In 2012, Morimoto *et al.* screened a library of ε-*N*-trifluoroacetyllysine-containing macrocylic peptides, ultimately identifying compound **14** (Fig. 6C), a highly potent and >10-fold selective SIRT2 inhibitor.[118] More recently, Huang *et al.* used macrocyclization to design compound **15** (Fig. 6C), a highly potent inhibitor, which was moderately selective toward SIRT1.[185] Interestingly, in this study, the authors saw that the type and length of the linker had large effects on potency and selectivity of the inhibitors.

Even though they do not strictly fall within the category of mechanism-based inhibitors, we highlight two examples of inhibitors, compounds **16** and **17** (Fig. 6C), which have been designed to occupy both the substrate and the NAD^+ binding pocket at the same time. In 2016, Wang *et al.* published compound **16**, which exhibited highly selective inhibition of SIRT1, a feature that is rarely seen in mechanism-based inhibitors. Although its potency is moderate, the selectivity can prove to be useful in terms of probing the role of SIRT1 in vivo. To date, there are very few mechanism-based inhibitors that show selective inhibition of SIRT3 over SIRT1 and SIRT2, and selectivity is often accompanied by poor potencies, for example, compound **5** (Fig. 6A).[184] In a recent report, Alhazzazi *et al.* published data on compound **17** (Fig. 6C), a SIRT3 inhibitor with similar selectivity as compound **5**, although showing slightly weaker inhibition.[187]

Interestingly, compound **17** is based on a glutamate- rather than a lysine motif, which might explain the low potency. Nevertheless, this compound was shown to increase acetylation of mitochondrial proteins and inhibit growth of head and neck squamous cell carcinoma cells.[187]

Progress has been made in the development of mechanism-based inhibition of SIRT1–3 and several potent and selective inhibitors have been identified. Although the efforts of several groups have generated important insight into the structure activity relationship of mechanism-based SIRT1–3 inhibitors, the lack of obvious trends makes it difficult to predict the potencies when designing such inhibitors.

3.2 Short-Chain ε-*N*-Acyllysine PTMs

3.2.1 ε-N-Formyllysine

ε-*N*-Formylation was first reported as lysine modification during a screening for acyl PTMs on histone 1,[37] but was extended to other nuclear proteins shortly after.[38] Further investigation showed that lysine formylation is promoted by oxidative DNA damage.[39] Despite the similarity of formyl and acetyl groups, formylated peptides proved to be poor substrates for sirtuins, showing only minor conversion by SIRT1, 3, and 6 in vitro.[125] Furthermore, the physiological relevance of this modification is still not fully understood.

3.2.2 ε-N-Propionyllysine and ε-N-Butyryllysine

The search for new acyl modifications started when a role for acetyl-CoA and acetyl transferases was demonstrated for the modification of lysine residues. As cell metabolism results in the formation of different acyl-CoAs, it was investigated whether these alternative thioesters could be substrates for acetyl transferases. This led to the identification of ε-*N*-propionyllysine (Kpro, Fig. 2B) and ε-*N*-butyryllysine (Kbut, Fig. 2B) on histone 4 and of propionyl-CoA and butyryl-CoA as substrates for p300 and CBP.[40] In terms of size and electrostatics, propionyl and butyryl acyl groups resemble the acetyl group (Fig. 2). Indeed, bromodomain proteins that bind to ε-*N*-acetyllysine, such as BRD4, BRD9, TAF1(2),[42] or CECR2,[43] can also bind to ε-*N*-propionyl- or ε-*N*-butyryllysine, respectively. Nevertheless, as the biochemical synthesis of acetyl-CoA, propionyl-CoA, and butyryl-CoA are distinct and depend on the energy state of the cell, a nonredundant function for these PTMs was suggested. However, a proteomic investigation of human cells is still missing for these PTMs, and only few Kpro/Kbut sites in mammals are known so far.[41,188] Butyrylation has recently been found on H4K5 and H4K8, which are known Kac sites linked to active transcription.

Unlike H4K5ac, butyrylation at H4K5 prevents the ability of Brdt, a bromodomain-containing transcriptional regulator, to recognize chromatin.[189] In addition to their discovery on histones, Kpro and Kbut have also been found on nonhistone proteins. Similar to acetyl-CoA,[40] propionyl-CoA has been shown to regulate the activity of propionyl synthetase, at least in bacteria.[190] In the same study,[190] depropionylase activity was demonstrated by the bacterial sirtuin CobB and (human) SIRT2 and SIRT3 in vitro. In a later study, depropionylase activity in cells was observed for SIRT1.[41] Further in vitro screenings for sirtuin activity showed appreciable deacylation of both Kpro and Kbut by SIRT1 and 2.[125,126]

3.2.3 ε-N-Crotonyllysine

ε-*N*-Crotonyllysine (Kcr, Fig. 2) was discovered in 2011 during a proteomic screening that led to the identification of 28 Kcr modification sites on human histones.[27] This modification was shown to be involved in transcriptional regulation, for example, in postmeiotically expressed genes linked to chromosome X[27] and in gene expression mediated by AF9 YEATS, a Kac/Kcr reader.[29,30] As demonstrated recently in a screening for Kcr readers, the bromodomain protein TAF1(2) can recognize this PTM as well. Even though the short carbon chain of the crotonyl moiety might suggest an overlap of regulatory functions with acetylation, Kac and Kcr sites on histones overlap only partially.[26,27] This partial overlap could be due to different sets of regulatory enzymes for the two modifications. Due to the conjugated double bond, both shape and electrostatic properties of Kcr are different from the other short-chain acyl groups. Underlining the importance of these differences Kbut and Kcr were not recognized by the same bromodomains.[43] Whole cell proteomics showed that ε-*N*-crotonyllysine residues are mainly found in nuclear proteins.[26] Although nonenzymatic acylation of lysine residues with crotonyl-CoA cannot be excluded, it has been shown that acetyl transferases p300 and CBP can catalyze lysine crotonylation.[26,54] Soon after the discovery of Kcr, HDAC3, SIRT1, and SIRT2 catalyzed hydrolysis of ε-*N*-crotonyllysine was shown in vitro.[126,191] In a different study, SIRT1–3 were shown to be the strongest binders of an H3K4cr peptide in a pull-down experiment.[28] A crystal structure of SIRT3 and the H3K4cr peptide showed that a phenylalanine residue (Phe180) was involved in π–π interaction with the conjugated double bond of the crotonyl group. This phenylalanine is conserved in SIRT1–3, and may explain the decrotonylase activity observed for these isotypes. However, SIRT3 was the only sirtuin enzyme that demonstrated decrotonylase activity in cells, based on the observation that only SIRT3 knockdown cells showed increased levels of crotonylation in histone proteins.[28]

3.2.4 Inhibitors Mimicking Short-Chain ε-N-Acyllysine PTMs

As inhibiting SIRT1–3 deacetylation have been of highest priority for researchers, selectively targeting the deacylation of these short-chain PTMs have gained little attention. Although crotonylation has been implicated in transcriptional regulation, the biological importance of propionyl and butyryl PTMs remains unclear. Developing compounds that could selectively inhibit the deacylation of these PTMs may further elucidate their relevance in the cell. Most likely, this is a difficult task to accomplish as mechanism-based inhibitors with short-chain acyl analogs often show inhibition of SIRT1–3 deacetylation.[115,120] Nevertheless, compounds **8** (Fig. 6B) and **18–21** (Fig. 7), which are Kpro or Kbut analogs, respectively, have not been evaluated for inhibition of short-chain deacylation.

3.3 Long-Chain ε-N-Acyllysine PTMs

3.3.1 ε-N-Myristoyllysine

Although the most abundant and best characterized PTMs involving long-chain acyl groups are cysteine palmitoylation and N-terminal glycine myristoylation, there are examples of long-chain acylated proteins that do not fall into these categories. This latter group includes interleukine 1α (IL-1α), TNF-α, the insulin receptor, and membrane immunoglobulin heavy chain, all presumably containing ε-N-myristoylated lysines (Kmyr, Fig. 2B).[44,45,192,193] TNF-α, a cytokine involved in the acute phase of inflammation, was shown to be myristoylated as early as 1992, and residues

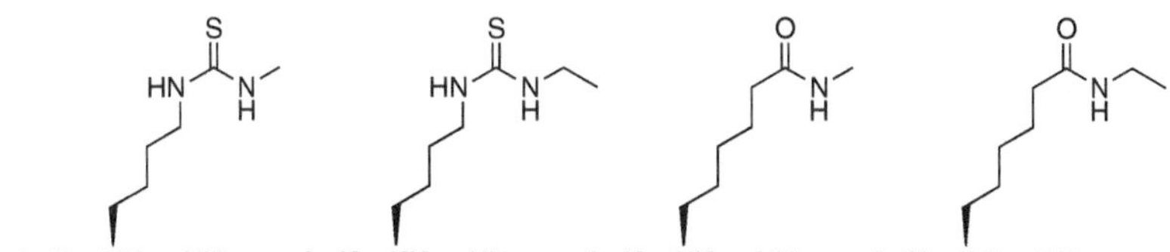

	Ac-**KacKKac**-NH$_2$ (**18**)	Ac-**KacKKac**-NH$_2$ (**19**)	Ac-**KacKKac**-NH$_2$ (**20**)	Ac-**KacKKac**-NH$_2$ (**21**)
SIRT1	5.6	83.4	149.8	118.2
SIRT2	1.65	ND	ND	77.9
SIRT3	2.2	ND	ND	32.3
SIRT5–7	ND	ND	ND	ND

Fig. 7 Sirtuin inhibitors with short acyl motifs; structures; and IC_{50} values (μM) for inhibition of deacetylation. **18,19**,[115] **20,21**.[120]

K19 and K20 were proposed to be the sites of modification (Kmyr, Fig. 2B).[44] It was then demonstrated that TNF-α production depends on SIRT6, even though the detailed mechanism was unknown.[194] These two observations were connected when it was proven that the release of TNFα was dependent on the hydrolysis of myristoylated K19/K20 by SIRT6.[20] R-Ras2, which is a protein in the GTPase family that is overexpressed in several tumor cell lines, has also been studied. Even though the specific acyl chain length was not addressed directly, long-chain acylation of R-Ras2 targets the protein toward the plasma membrane where it interacts with other proteins and eventually leads to increased cell proliferation.[46]

Long-chain acyl groups are removed by SIRT6 that features a broad lipophilic pocket in the catalytic site. Though SIRT6 was the first sirtuin proven to exhibit fatty deacylase activity in cells, long-chain acyl groups are also cleaved by SIRT1–3, as demonstrated in vitro for SIRT1–3[125–127] and for SIRT2 in cells.[195] Crystallographic data reveal how SIRT1–3 accommodate long-chain acyl groups. Despite different residues in the acyl pocket of SIRT1–3 and SIRT6, several hydrophobic residues create a favorable environment for the accommodation of a long carbon chain, indicating a similar mode of interaction for SIRT1–3 and SIRT6.[196]

Additionally, RNA-induced SIRT7 deacylase activity toward long-chain acylated lysine residues has recently been described.[129] This is intriguing since SIRT7—like SIRT6—is a class IV sirtuin, and that it has been shown to be related to RNA metabolism,[170–172] and as SIRT7-mediated deacetylation was reported to be activated by DNA strands.[130]

3.3.2 Inhibitors Mimicking Long-Chain Acyllysine PTMs

Much like the short-chain acyl groups, the biological importance of long-chain acyl PTMs has yet to be fully understood. To probe their relevance, sirtuin inhibitors could prove to be important tool compounds. Several mechanism-based inhibitors with acyl analogs of various lengths have been reported, such as compounds **9**, **10**, and **13** (Fig. 6B) and compounds **22–24** (Fig. 8). However, some of these compounds show cross-inhibition against SIRT1–3 and SIRT6 deacetylation, as exemplified by compounds **22** and **24** (Fig. 8). Furthermore, these inhibitors often show different potencies against SIRT6 deacetylation and demyristoylation as seen for compound **24** (Fig. 8).[197] More importantly, most long-chain analog inhibitors have not been evaluated against SIRT1–3 long-chain deacylation. The highly selective SIRT2 inhibitor **10** (Fig. 6B) is one example, which was shown to be highly potent against SIRT2 deacetylation and only weakly potent

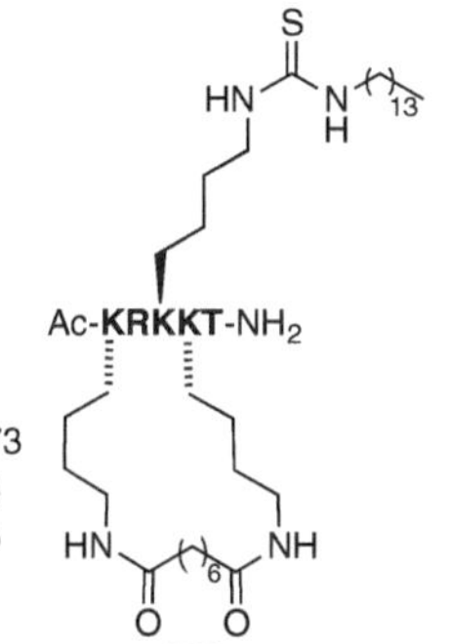

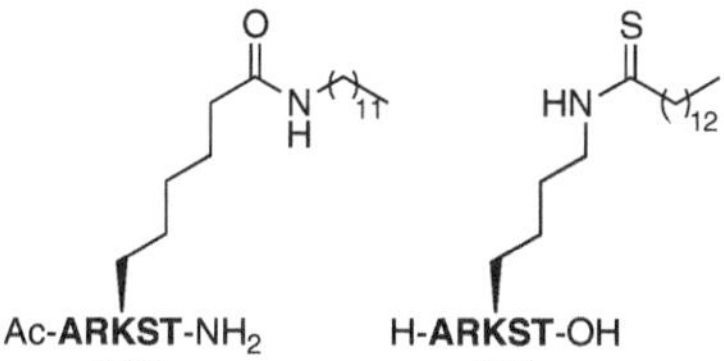

	(22)	(23)	(24)
SIRT1	0.73	ND	4.4
SIRT2	6.4	ND	2.6
SIRT3	3.5	ND	5.6
SIRT5	>300	ND	ND
SIRT6 (Kac)	ND	ND	8.2
SIRT6 (Kmyr)	0.32	20.9	0.32

Fig. 8 Sirtuin inhibitors with long acyl motifs; structures and IC_{50} values (μM) for inhibition of deacetylation (SIRT1–3), desuccinylation (SIRT5), and deacetylation and demyristoylation (SIRT6, Kac, and Kmyr, respectively). **22**,[116] **23**,[120] **24**[197].

against SIRT6 demyristoylation, but the inhibitory effect on SIRT1–3 long-chain deacylation was not assessed.[182]

3.4 Negatively Charged ε-*N*-Acyllysine PTMs

3.4.1 ε-N-Malonyl-, ε-N-Succinyl-, and ε-N-Glutaryllysine

ε-*N*-Malonyl-, ε-*N*-succinyl-, and ε-*N*-glutaryllysine represent a distinct group of lysine PTMs (Kmal, Ksuc, and Kglu, respectively; Fig. 2A). Posttranslational modification of lysine with these carboxyacyl groups leads to a change in the overall charge from +1 to −1. For this reason, it is likely that these PTMs will affect protein structure and interaction with other proteins more than smaller and noncharged acyl groups. ε-*N*-Succinyllysine was the first carboxyacyl group to be recognized as a PTM.[18] Malonyl was discovered shortly after,[17] followed by glutaryl.[19] Independently, two groups reported SIRT5 as the enzyme responsible for the deacylation of Kmal and Ksuc,[15,17] and later the deglutarylation activity of this enzyme was also demonstated.[19] Importantly, SIRT5 was shown to have a preference for these PTMs over acetyl, due to the presence of two specific residues at the end of the acyl binding pocket, an arginine (Arg105) and a tyrosine (Tyr102). These residues effectively coordinate the negatively charged carboxylate moiety of the substrate and thus create a strong interaction with the acyl group.[15] SIRT5 mediated desuccinylation has been linked to activation of enzymes, such as 3-hydroxy-3-methylglutaryl-CoA synthase 2, the rate-limiting enzyme in ketone body synthesis,[198] and respiratory chain complexes.[199] However, it has also been shown to negatively regulate the pyruvate dehydrogenase complex and succinate dehydrogenase.[200] Malonylation of lysine residues was shown to be involved in the regulation of protein activity as well.[33] In general, proteins regulated by the demalonylase activity of SIRT5 are part of gluconeogenesis, glycolysis, and the urea cycle. Among these proteins, aldolase B and glyceraldehyde 3-phosphate dehydrogenase (GAPDH) are affected by SIRT5.[201] Altered Kmal levels have been described in malonyl-CoA decarboxylase (MCD) deficient cells, an inborn metabolic disorder, which results in increased levels of malonyl-CoA in cells.[31] When the stoichiometry of Kmal was calculated, it turned out that enzymes involved in fatty acid metabolism, such as acetyl-CoA acetyltransferase 1 (ACAT1) and very long-chain acyl-CoA dehydrogenase (VLCAD) showed relevant differences between wild type and MCD deficient cells. Malonylation levels of two sites on VLCAD and one site on ACAT1 were shown to be more than 100-fold higher in MCD deficient cells than in wild type cells, suggesting a substantial regulation of this protein by malonylation.[31] Analysis of the sequences where these modifications have been found showed that aliphatic amino acids, such as Gly and Ala, are overrepresented close to Kmal and Ksuc residues.[31,200] In general, Kmal sites overlap only partially with Kac and Ksuc, suggesting a distinct regulation performed by Kmal residues compared with other acyllysines. When Kmal and Ksuc substrates were screened for cellular localization in human fibroblasts and mouse liver cells, respectively, it was found that succinylated proteins were almost absent in cytoplasm,[198] while malonylated substrates are approximately equally present in nucleus, cytoplasm, and mitochondria.[31] This is interesting, given that SIRT5 is the regulatory enzyme for this PTM and is manly localized in mitochondria. Unlike malonyl and succinyl, less is known about the function of ε-*N*-glutarylated lysine residues, but it appears that this PTM is found on several proteins as well.[19] One example is carbamoyl phosphate synthase 1 (CPS1),[19] a protein which has also been shown to be regulated by acetylation,[201] malonylation,[202] and succinylation.[15] Although these three carboxyacyl PTMs are similar, they are part of different biological pathways. It could be rationalized that a cell could install one or the other to regulate proteins, depending on the availability of the corresponding acyl-CoAs.

3.4.2 SIRT5 as Deacylase

SIRT5 activity has been implicated in diseases, such as lung cancer; and hence inhibition of SIRT5 has also received attention over the last few years. SIRT5-mediated desuccinylation of superoxide dismutase 1 (SOD1), which protects the cell from oxidative stress, increases its activity.[203] SOD1 has been implicated to promote tumor growth in nonsmall cell lung cancer (NSCLC), and inhibition of its activity has been shown to decrease growth of NSCLC.[204,205] Moreover, desuccinylation of another SIRT5 substrate, pyruvate kinase M2 (PKM2) leads to increased activity and promotes tumor growth.[206] These examples illustrate some of the therapeutic potential of inhibiting SIRT5.

3.4.3 SIRT4 and SIRT7 as Deacylases

Recently, other negatively charged PTMs—hydroxy-3-methylglutaryl (Khmg), 3-methylglutaryl (Kmg), and 3-methylglutaconyl (Kmgc), were discovered (Fig. 2). Enzymes involved in leucine metabolism, such as methylcrotonyl-CoA carboxylase (MCCC), are affected by these new PTMs. Acylation of MCCC leads to impaired complex formation and function. Consequently, leucine catabolism is reduced, which promotes insulin secretion in pancreatic islets.[35] Although Kmg and Khmg have been shown to bind to SIRT5 with relatively high affinity, they are primarily deacylated by SIRT4.[35,207] Moreover, SIRT7 has recently been described as a desuccinylase as well. SIRT7-mediated desuccinylation of H3K122 was shown to be involved in DNA double-strand breaks (DSBs) repair.[131] Although this new deacylase activity requires further investigation, SIRT7 could be the sirtuin that removes negatively charged ε-*N*-acyllysine residues in the cell nuclei.

3.4.4 Inhibitors Mimicking Negatively Charged ε-N-Acyllysine PTMs

Over the last few years, several mechanism-based SIRT5 inhibitors have been reported. For SIRT5 inhibitors, selectivity has been more easily achieved, due to its preference for carboxyacyllysines.[208] It has been shown that glutaryl- and succinyllysine bind to SIRT5 with higher affinity compared to malonyllysine.[19,207] One explanation for this could be that the malonyl moiety cannot reach the crucial arginine and tyrosine residues in the binding pocket as efficiently as succinyl- and glutaryllysines. In addition, acyl groups containing a terminal carboxy motif with longer chain lengths than glutaryl have been shown to have decreased binding affinity and are converted much slower.[19,207] Most probably this is because the longer chains cannot be accommodated in the binding pocket while retaining the Arg and Tyr interactions and/or a productive orientation of the amide carbonyl oxygen. Consequently, SIRT5 inhibitors are most often designed with succinyl- and glutaryl analogs, which show similar potencies. By utilizing analogs of these acyl motifs, in combination with various amide isosteres, several selective mechanism-based SIRT5 inhibitors have been synthesized.[114,115,117,209] Evident by comparison of compounds **27–32** (Fig. 9),

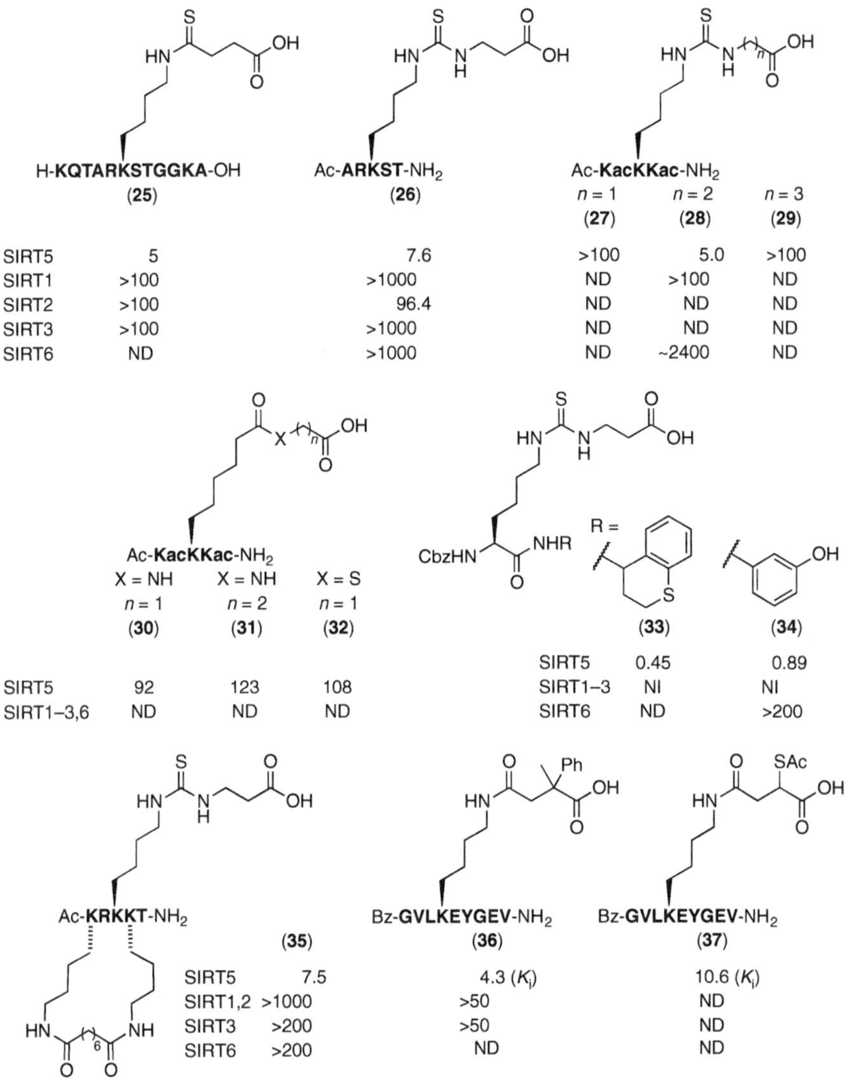

Fig. 9 Sirtuin inhibitors with carboxyacyl motifs; structures and IC_{50} values (μM) for inhibition of desuccinylation (SIRT5), deacetylation (SIRT1–3), and demyristoylation (SIRT6). **25**,[209] **26**,**35**[117] **27-29**,[115] **30-32**,[120] **33**,**34**,[114] **36**,**37**[207].

thioamide and thiourea isosteres possess significantly higher potencies compared to L-Asu isosteres.[115,120] This is in agreement with other L-Asu-based inhibitors showing lower potencies compared to their thioamide or thiourea counterparts.[113,116,119,120,179] Similar to SIRT1–3 inhibitors, peptidic scaffolds of different lengths, macrocycles, and pseudopeptide scaffolds have been utilized in an attempt to improve the potency of SIRT5 inhibitors. As seen with Kac-mimicking inhibitors of SIRT1–3, little difference can be seen when varying the peptide scaffold for these SIRT5 inhibitors (Fig. 9). However, the choice of amide isostere in combination with acyl chain length appears to be highly important. A study by Zang *etal.* showed that a thiourea-based succinyl analog, compound **27** (Fig. 9), was significantly less potent than the corresponding thiourea-based glutaryl analog, compound **28** (Fig. 9). On the other hand, the thioamide-based succinyl analog **25** (Fig. 9) exhibits inhibition of SIRT5 similar to that of compound **28**. This might indicate that the binding affinity of the thiourea-based succinyl analogs is impaired, and perhaps they cannot adopt a desirable orientation to interact with the crucial arginine and tyrosine residues. Similarly, compound **29** (Fig. 9) shows weak inhibition presumably due to a nonoptimal chain length. Thus, when using thiourea isosteres, glutaryl analogs seem to be the ideal choice for inhibiting SIRT5.

Although these compounds often show high selectivity toward SIRT5, there are only few compounds, such as thioureas **33** and **34**, that have been reported to have submicromolar IC_{50} values (Fig. 9). The most potent SIRT5 inhibitor reported to date, compound **33**, with an IC_{50} value of 0.45 μM, is based on a pseudopeptidic scaffold (Fig. 9).[114] Interestingly, an ethylester prodrug of this compound was shown to inhibit growth of MCF-7 cells.[114] Compounds **36** and **37** (Fig. 9) are examples of SIRT5 inhibitors that inhibit binding of both substrate and cosubstrate, much like compounds **16** and **17** (Fig. 6C). Utilizing the carboxyacyl preference of SIRT5, these compounds were designed as ε-*N*-succinyllysine analogs carrying bulky residues to simultaneously target the binding site of NAD^+.[207] Although these inhibitors exhibit 10-fold lower inhibition than, for example, compound **33**, compound **36** shows high selectivity toward SIRT5, which warrants further studies of this type of inhibitor to potentially improve potency.

The utility of these SIRT5 inhibitors to be applied in biological settings is, however, impaired by the fact that they contain free carboxylic acids, which are known to impair cell permeability.[210] While conventional prodrug strategies could possibly circumvent this issue, carboxylic acid bioisosteres should be considered as alternatives for future development of SIRT5 inhibitors. Additionally, as SIRT5 is mainly located in the mitochondria, targeting this compartment will be another issue to overcome for in vivo application of SIRT5 inhibitors.

It is important to note that in addition to Kmg, Khmg, and Kmgc SIRT4 has a preference for glutaryllysine.[35] This makes the selectivity of glutaryl analog inhibitors for SIRT5 uncertain, and it suggests that these inhibitors could inhibit SIRT4 as well. However, the absence of commercially available active SIRT4 is a major hurdle when probing the selectivity of the glutaryl type inhibitors.

3.5 Other ε-*N*-Acyllysine Modifications

ε-*N*-(3-Phosphoglyceryl)lysine (Kpg, Fig. 2A) was discovered when 1,3-biphosphoglycerate (1,3-BPG) was investigated as potential acyl donor for Kpg.This PTM was initially discovered on GAPDH, the enzyme which converts 3-phosphoglyceraldehyde into 1,3-BPG. Subsequent cell lysate proteomics was used to identify Kpg on several other proteins as well.[36] Presumably reflecting the increased glycolytic flux, Kpg levels were shown to be increased when cells were incubated with high concentrations of glucose. Interestingly, no enzyme has been shown to deacylate this modification, or its dephosphorylated analog, thus far.

Biotin, also known as vitamin B7, is a coenzyme that mediates the transfer of carbon dioxide for five carboxylases (acetyl-CoA carboxylases 1 and 2, β-methylcrotonyl-CoA-, propionyl-CoA-, and pyruvate carboxylase). Several lysine residues of histone proteins have been shown to be modified with biotin (Kbio, Fig. 2B) and nucleosomal studies demonstrated that biotinylation, at least in some sites, induces chromatin condensation and in turn repression of transcription.[48,211] Despite the interesting role of biotinylated lysines in histones, no nuclear debiotinylase has been identified so far. However, weak debiotinylase activity has been shown for SIRT4 in vitro.[49]

Lipoic acid is a fundamental cofactor involved in metabolism, and is found as ε-*N*-lipoyllysine (Klip, Fig. 2B).[47] It is known that cells can salvage exogenous lipoic acid to survive,[212,213] but when the first mammalian lipoic acid synthase was cloned, a complementary de novo synthesis of lipoic acid in mammals was proposed.[214] Interestingly, SIRT4 was recently described as a lipoamidase that targets the pyruvate dehydrogenase complex and down-regulates its activity.[49] This was in contrast to previous substrate screenings that failed to detect significant delipoylase activity for SIRT4,[125,126] likely due to the use of nonactive recombinant SIRT4.

ε-*N*-(2-Hydroxyisobutyryl)lysine (Khib, Fig. 2B) was discovered during a proteomic screening of histone modifications from mouse testis cells.[50] In this study, H4K8hib was linked to enhanced gene expression in postmeiotic cells, a process during which Kcr and Kac were reported to play a role as well.[27,215] Furthermore, preliminary results have demonstrated that HDAC1–3 can hydrolyze Khib in vitro.[50]

ε-*N*-(β-Hydroxybutyryl)lysine (Kbhb, Fig. 2B) was discovered using the same approach as for Khib.[51] As β-hydroxybutyrate is one of the ketone bodies exploited for energy production when glucose levels are low, it was not surprising that Kbhb levels accumulated in mice upon fasting. β-Hydroxybutyrylation of lysine is associated with enhanced gene expression, in a way similar to the transcriptional activation induced by H3K4 trimethylation and H3K9ac. As β-hydroxybutyrate and ketone bodies have been proposed to be beneficial in several diseases, such as epilepsy[216] and cancer,[217] it would be interesting to determine whether there is a link between these positive effects and an increase in lysine β-hydroxybutyrylation.

It is well known that peroxidation of lipids leads to the generation of reactive species, which are detrimental for cells.[218,219] Among the electrophiles generated in this process, 4-oxo-2-nonenal (4-one) represents a major product,[220] which is toxic in neuroblastoma cells.[221] The reactive 4-one was shown to form adducts with histones, including ε-*N*-4-oxo-nonenoyllysine (K4one, Fig. 2B), that are able to disrupt nucleosome assembly.[52] Recently, this modification can be hydrolyzed by SIRT2.[53] Further studies are needed, but these results suggest that SIRT2 has a role in oxidative stress regulation.

Although several of these ε-*N*-acyllysine PTMs have not been shown to be deacylated by sirtuins, based on their structure it could be speculated that they are sirtuin substrates. The shorter Khib and Kbhb could possibly be recognized and deacylated by SIRT1–3, 6, or 7. Moreover, the terminal phosphate ester moiety of Kpg could be considered a carboxylic acid bioisostere, which might be recognized by SIRT4 and SIRT5. So far no sirtuin inhibitors have been designed to target the deacylation of these PTMs. This uncharted area could be interesting to explore, since they could provide more insight into the biological role of these PTMs and their relation to the sirtuin enzymes.

4. CONCLUSIONS AND PERSPECTIVES

Since their discovery, our understanding of the sirtuins has increased immensely. The numerous substrates that have been identified illustrate the important roles of sirtuins in crucial cellular processes, including gene regulation and metabolism.

Insight into the mechanism of sirtuin-catalyzed deacylation has led to the development of various amide analogs, which have been utilized in the design of potent mechanism-based sirtuin inhibitors. Furthermore, identification of acyl preferences and structural information has guided researchers to develop isotype-specific inhibitors. The continued emergence of evidence that implicate sirtuins in various diseases calls for the development of additional potent and selective sirtuin inhibitors. Such compounds may be used as tools to further study the sirtuins and may serve as lead compounds for drug development. The collective effort of several groups has led to remarkable progress within the field over the last 2 decades. Nonetheless, there is still significant room for further improvement of the arsenal of sirtuin inhibitors available.

One strategy for achieving inhibitors with a degree of isotype selectivity, as well as potency, is to rely on a combination of mechanistic insight and information from substrate selectivity. In particular, taking advantage of the preferences of individual sirtuin isotypes for different ε-*N*-acyllysine PTMs have proved fruitful for achieving subtype selective inhibitors. Interestingly, the metabolomes of cells contain a variety of acyl-CoA species and since acyl-CoAs are recognized as the major donors for ε-*N*-acyllysine modifications, yet unknown PTMs targeted by sirtuins may still exist. Investigation of sirtuin substrate specificity and utilizing putative new catalytic insight for generation of future inhibitors holds promise. Finally, efforts to transform the inherently, highly peptide-inspired mechanism-based inhibitors into more "drug-like" molecules will be an interesting future challenge.

REFERENCES

1. Ivy JM, Klar AJS, Hicks JB. Cloning and characterization of four *SIR* genes of *Saccharomyces cerevisiae*. *Mol Cell Biol.* 1986;6(2):688–702.
2. Rine J, Herskowitz I. Four genes responsible for a position effect on expression from *HML* and *HMR* in *Saccharomyces cerevisiae*. *Genetics.* 1987;116(1):9–22.
3. Frye RA. Phylogenetic classification of prokaryotic and eukaryotic Sir2-like proteins. *Biochem Biophys Res Commun.* 2000;273(2):793–798.
4. Marmorstein R. Structure of histone deacetylases: insights into substrate recognition and catalysis. *Structure.* 2001;9(12):1127–1133.
5. Michishita E, Park JY, Burneskis JM, Barrett JC, Horikawa I. Evolutionarily conserved and nonconserved cellular localizations and functions of human SIRT proteins. *Mol Biol Cell.* 2005;16(10):4623–4635.
6. Vaquero A, Scher MB, Lee DH, Sutton A, Cheng H-L, Alt FW, Serrano L, Sternglanz R, Reinberg D. SirT2 is a histone deacetylase with preference for histone H4 Lys 16 during mitosis. *Genes Dev.* 2006;20(10):1256–1261.
7. Tanno M, Sakamoto J, Miura T, Shimamoto K, Horio Y. Nucleocytoplasmic shuttling of the NAD^+-dependent histone deacetylase SIRT1. *J Biol Chem.* 2007;282(9):6823–6832.
8. Matsushita N, Yonashiro R, Ogata Y, Sugiura A, Nagashima S, Fukuda T, Inatome R, Yanagi S. Distinct regulation of mitochondrial localization and stability of two human Sirt5 isoforms. *Genes Cells.* 2011;16(2):190–202.

9. Tanny JC, Dowd GJ, Huang J, Hilz H, Moazed D. An enzymatic activity in the yeast Sir2 protein that is essential for gene silencing. *Cell.* 1999;99(7):735–745.
10. Frye RA. Characterization of five human cDNAs with homology to the yeast SIR2 gene: Sir2-like proteins (sirtuins) metabolize NAD and may have protein ADP-ribosyltransferase activity. *Biochem Biophys Res Commun.* 1999;260(1):273–279.
11. Landry J, Sutton A, Tafrov ST, Heller RC, Stebbins J, Pillus L, Sternglanz R. The silencing protein SIR2 and its homologs are NAD-dependent protein deacetylases. *Proc Natl Acad Sci USA.* 2000;97(11):5807–5811.
12. Imai S-I, Armstrong CM, Kaeberlein M, Guarente L. Transcriptional silencing and longevity protein Sir2 is an NAD-dependent histone deacetylase. *Nature.* 2000;403 (6771):795–800.
13. Smith JS, Brachmann CB, Celic I, Kenna MA, Muhammad S, Starai VJ, Avalos JL, Escalante-Semerena JC, Grubmeyer C, Wolberger C, Boeke JD. A phylogenetically conserved NAD^+-dependent protein deacetylase activity in the Sir2 protein family. *Proc Natl Acad Sci USA.* 2000;97(12):6658–6663.
14. Finnin MS, Donigian JR, Pavletich NP. Structure of the histone deacetylase SIRT2. *Nat Struct Mol Biol.* 2001;8(7):621–625.
15. Du JT, Zhou YY, Su XY, Yu JJ, Khan S, Jiang H, Kim J, Woo J, Kim JH, Choi BH, He B, Chen W, Zhang S, Cerione RA, Auwerx J, Hao Q, Lin HN. Sirt5 is a NAD-dependent protein lysine demalonylase and desuccinylase. *Science.* 2011;334(6057):806–809.
16. Haigis MC, Mostoslavsky R, Haigis KM, Fahie K, Christodoulou DC, Murphy AJ, Valenzuela DM, Yancopoulos GD, Karow M, Blander G, Wolberger C, Prolla TA, Weindruch R, Alt FW, Guarente L. SIRT4 inhibits glutamate dehydrogenase and opposes the effects of calorie restriction in pancreatic β cells. *Cell.* 2006;126 (5):941–954.
17. Peng C, Lu Z, Xie Z, Cheng Z, Chen Y, Tan M, Luo H, Zhang Y, He W, Yang K, Zwaans BMM, Tishkoff D, Ho L, Lombard D, He TC, Dai J, Verdin E, Ye Y, Zhao Y. The first identification of lysine malonylation substrates and its regulatory enzyme. *Mol Cell Proteomics.* 2011;10(12). 10.1074/mcp.M111.012658.
18. Zhang Z, Tan M, Xie Z, Dai L, Chen Y, Zhao Y. Identification of lysine succinylation as a new post-translational modification. *Nat Chem Biol.* 2011;7(1):58–63.
19. Tan M, Peng C, Anderson KA, Chhoy P, Xie Z, Dai L, Park J, Chen Y, Huang H, Zhang Y, Ro J, Wagner GR, Green MF, Madsen AS, Schmiesing J, Peterson BS, Xu G, Ilkayeva OR, Muehlbauer MJ, Braulke T, Mühlhausen C, Backos DS, Olsen CA, McGuire PJ, Pletcher SD, Lombard DB, Hirschey MD, Zhao Y. Lysine glutarylation is a protein posttranslational modification regulated by SIRT5. *Cell Metab.* 2014;19(4):605–617.
20. Jiang H, Khan S, Wang Y, Charron G, He B, Sebastian C, Du J, Kim R, Ge E, Mostoslavsky R, Hang HC, Hao Q, Lin H. SIRT6 regulates TNF-α secretion through hydrolysis of long-chain fatty acyl lysine. *Nature.* 2013;496(7443):110–113.
21. Choudhary C, Kumar C, Gnad F, Nielsen ML, Rehman M, Walther TC, Olsen JV, Mann M. Lysine acetylation targets protein complexes and co-regulates major cellular functions. *Science.* 2009;325(5942):834–840.
22. Kim SC, Sprung R, Chen Y, Xu Y, Ball H, Pei J, Cheng T, Kho Y, Xiao H, Xiao L, Grishin NV, White M, Yang XJ, Zhao Y. Substrate and functional diversity of lysine acetylation revealed by a proteomics survey. *Mol Cell.* 2006;23(4):607–618.
23. Zhao S, Xu W, Jiang W, Yu W, Lin Y, Zhang T, Yao J, Zhou L, Zeng Y, Li H, Li Y, Shi J, An W, Hancock SM, He F, Qin L, Chin J, Yang P, Chen X, Lei Q, Xiong Y, Guan KL. Regulation of cellular metabolism by protein lysine acetylation. *Science.* 2010;327 (5968):1000–1004.
24. Choudhary C, Weinert BT, Nishida Y, Verdin E, Mann M. The growing landscape of lysine acetylation links metabolism and cell signalling. *Nat Rev Mol Cell Biol.* 2014;15 (8):536–550.

25. Schölz C, Weinert BT, Wagner SA, Beli P, Miyake Y, Qi J, Jensen LJ, Streicher W, McCarthy AR, Westwood NJ, Lain S, Cox J, Matthias P, Mann M, Bradner JE, Choudhary C. Acetylation site specificities of lysine deacetylase inhibitors in human cells. *Nat Biotechnol.* 2015;33(4):415–423.
26. Wei W, Mao A, Tang B, Zeng Q, Gao S, Liu X, Lu L, Li W, Du JX, Li J, Wong J, Liao L. Large-scale identification of protein crotonylation reveals its role in multiple cellular functions. *J Proteome Res.* 2017;16(4):1743–1752.
27. Tan M, Luo H, Lee S, Jin F, Yang JS, Montellier E, Buchou T, Cheng Z, Rousseaux S, Rajagopal N, Lu Z, Ye Z, Zhu Q, Wysocka J, Ye Y, Khochbin S, Ren B, Zhao Y. Identification of 67 histone marks and histone lysine crotonylation as a new type of histone modification. *Cell.* 2011;146(6):1016–1028.
28. Bao X, Wang Y, Li X, Li XM, Liu Z, Yang T, Wong CF, Zhang J, Hao Q, Li XD. Identification of "erasers" for lysine crotonylated histone marks using a chemical proteomics approach. *eLife.* 2014;3:e02999.
29. Li Y, Sabari BR, Panchenko T, Wen H, Zhao D, Guan H, Wan L, Huang H, Tang Z, Zhao Y, Roeder RG, Shi X, Allis CD, Li H. Molecular coupling of histone crotonylation and active transcription by AF9 YEATS domain. *Mol Cell.* 2016;62(2):181–193.
30. Andrews FH, Shinsky SA, Shanle EK, Bridgers JB, Gest A, Tsun IK, Krajewski K, Shi X, Strahl BD, Kutateladze TG. The Taf14 YEATS domain is a reader of histone crotonylation. *Nat Chem Biol.* 2016;12(6):396–398.
31. Colak G, Pougovkina O, Dai L, Tan M, te Brinke H, Huang H, Cheng Z, Park J, Wan X, Liu X, Yue WW, Wanders RJA, Locasale JW, Lombard DB, de Boer VCJ, Zhao Y. Proteomic and biochemical studies of lysine malonylation suggest its malonic aciduria-associated regulatory role in mitochondrial function and fatty acid oxidation. *Mol Cell Proteomics.* 2015;14(11):3056–3071.
32. Xie Z, Dai J, Dai L, Tan M, Cheng Z, Wu Y, Boeke JD, Zhao Y. Lysine succinylation and lysine malonylation in histones. *Mol Cell Proteomics.* 2012;11(5):100–107.
33. Hirschey MD, Zhao Y. Metabolic regulation by lysine malonylation, succinylation, and glutarylation. *Mol Cell Proteomics.* 2015;14(9):2308–2315.
34. Weinert BT, Schölz C, Wagner SA, Iesmantavicius V, Su D, Daniel JA, Choudhary C. Lysine succinylation is a frequently occurring modification in prokaryotes and eukaryotes and extensively overlaps with acetylation. *Cell Rep.* 2013;4(4):842–851.
35. Anderson KA, Huynh FK, Fisher-Wellman K, Stuart JD, Peterson BS, Douros JD, Wagner GR, Thompson JW, Madsen AS, Green MF, Sivley RM, Ilkayeva OR, Stevens RD, Backos DS, Capra JA, Olsen CA, Campbell JE, Muoio DM, Grimsrud PA, Hirschey MD. SIRT4 is a lysine deacylase that controls leucine metabolism and insulin secretion. *Cell Metab.* 2017;25(4):838–855.
36. Moellering RE, Cravatt BF. Functional lysine modification by an intrinsically reactive primary glycolytic metabolite. *Science.* 2013;341(6145):549–553.
37. Wiśniewski JR, Zougman A, Krüger S, Mann M. Mass spectrometric mapping of linker histone H1 variants reveals multiple acetylations, methylations, and phosphorylation as well as differences between cell culture and tissue. *Mol Cell Proteomics.* 2007;6(1):72–87.
38. Wiśniewski JR, Zougman A, Mann M. N^{ε}-Formylation of lysine is a widespread post-translational modification of nuclear proteins occurring at residues involved in regulation of chromatin function. *Nucleic Acids Res.* 2007;36(2):570–577.
39. Jiang T, Zhou XF, Taghizadeh K, Dong M, Dedon PC. *N*-formylation of lysine in histone proteins as a secondary modification arising from oxidative DNA damage. *Proc Natl Acad Sci USA.* 2007;104(1):60–65.
40. Chen Y, Sprung R, Tang Y, Ball H, Sangras B, Kim SC, Falck JR, Peng J, Gu W, Zhao Y. Lysine propionylation and butyrylation are novel post-translational modifications in histones. *Mol Cell Proteomics.* 2007;6(5):812–819.

41. Cheng Z, Tang Y, Chen Y, Kim S, Liu H, Li SSC, Gu W, Zhao Y. Molecular characterization of propionyllysines in non-histone proteins. *Mol Cell Proteomics.* 2009;8(1):45–52.
42. Vollmuth F, Geyer M. Interaction of propionylated and butyrylated histone H3 lysine marks with Brd4 bromodomains. *Angew Chem Int Ed.* 2010;49(38):6768–6772.
43. Flynn EM, Huang OW, Poy F, Oppikofer M, Bellon SF, Tang Y, Cochran AG. A subset of human bromodomains recognizes butyryllysine and crotonyllysine histone peptide modifications. *Structure.* 2015;23(10):1801–1814.
44. Stevenson FT, Bursten SL, Locksley RM, Lovett DH. Myristyl acylation of the tumor necrosis factor α precursor on specific lysine residues. *J Exp Med.* 1992;176(4): 1053–1062.
45. Stevenson FT, Bursten SL, Fanton C, Locksley RM, Lovett DH. The 31-kDa precursor of interleukin la is myristoylated on specific lysines within the 16-kDa N-terminal propiece. *Proc Natl Acad Sci USA.* 1993;90:7245–7249.
46. Zhang X, Spiegelman NA, Nelson OD, Jing H, Lin H. SIRT6 regulates Ras-related protein R-Ras2 by lysine defatty-acylation. *eLife.* 2017;6:e25158.
47. Perham RN. Swinging arms and swinging domains in multifunctional enzymes: catalytic machines for multistep reactions. *Annu Rev Biochem.* 2000;69(1): 961–1004.
48. Filenko NA, Kolar C, West JT, Smith SA, Hassan YI, Borgstahl GEO, Zempleni J, Lyubchenko YL. The role of histone H4 biotinylation in the structure of nucleosomes. *PLoS One.* 2011;6(1):e16299.
49. Mathias RA, Greco TM, Oberstein A, Budayeva HG, Chakrabarti R, Rowland EA, Kang Y, Shenk T, Cristea IM. Sirtuin 4 is a lipoamidase regulating pyruvate dehydrogenase complex activity. *Cell.* 2014;159(7):1615–1625.
50. Dai L, Peng C, Montellier E, Lu Z, Chen Y, Ishii H, Debernardi A, Buchou T, Rousseaux S, Jin F, Sabari BR, Deng Z, Allis CD, Ren B, Khochbin S, Zhao Y. Lysine 2-hydroxyisobutyrylation is a widely distributed active histone mark. *Nat Chem Biol.* 2014;10(5):365–370.
51. Xie Z, Zhang D, Chung D, Tang Z, Huang H, Dai L, Qi S, Li J, Colak G, Chen Y, Xia C, Peng C, Ruan H, Kirkey M, Wang D, Jensen LM, Kwon OK, Lee S, Pletcher SD, Tan M, Lombard DB, White KP, Zhao H, Li J, Roeder RG, Yang X, Zhao Y. Metabolic regulation of gene expression by histone lysine β-hydroxybutyrylation. *Mol Cell.* 2016;62(2):194–206.
52. Galligan JJ, Rose KL, Beavers WN, Hill S, Tallman KA, Tansey WP, Marnett LJ. Stable histone adduction by 4-oxo-2-nonenal: a potential link between oxidative stress and epigenetics. *J Am Chem Soc.* 2014;136(34):11864–11866.
53. Jin J, He B, Zhang X, Lin H, Wang Y. SIRT2 reverses 4-oxononanoyl lysine modification on histones. *J Am Chem Soc.* 2016;138(38):12304–12307.
54. Sabari BR, Tang Z, Huang H, Yong-Gonzalez V, Molina H, Kong HE, Dai L, Shimada M, Cross JR, Zhao Y, Roeder RG, Allis CD. Intracellular crotonyl-CoA stimulates transcription through p300-catalyzed histone crotonylation. *Mol Cell.* 2015;58(2): 203–215.
55. Yuan H, Marmorstein R. Histone acetyltransferases: rising ancient counterparts to protein kinases. *Biopolymers.* 2013;99(2):98–111.
56. Simic Z, Weiwad M, Schierhorn A, Steegborn C, Schutkowski M. The ε-amino group of protein lysine residues is highly susceptible to nonenzymatic acylation by several physiological acyl-CoA thioesters. *ChemBioChem.* 2015;16(16):2337–2347.
57. Wagner GR, Bhatt DP, O'Connell TM, Thompson JW, Dubois LG, Backos DS, Yang H, Mitchell GA, Ilkayeva OR, Stevens RD, Grimsrud PA, Hirschey MD. A class of reactive acyl-CoA species reveals the non-enzymatic origins of protein acylation. *Cell Metab.* 2017;25(4):823–837.

58. James AM, Hoogewijs K, Logan A, Hall AR, Ding S, Fearnley IM, Murphy MP. Non-enzymatic *N*-acetylation of lysine residues by acetylCoA often occurs via a proximal S-acetylated thiol intermediate sensitive to glyoxalase II. *Cell Rep.* 2017;18(9):2105–2112.
59. Wang Q, Zhang Y, Yang C, Xiong H, Lin Y, Yao J, Li H, Xie L, Zhao W, Yao Y, Ning ZB, Zeng R, Xiong Y, Guan KL, Zhao S, Zhao GP. Acetylation of metabolic enzymes coordinates carbon source utilization and metabolic flux. *Science.* 2010;327(5968): 1004–1007.
60. Lin Y-y. Lu J-y. Zhang JM, Walter W, Dang W, Wan J, Tao S-c. Qian J, Zhao Y, Boeke JD, Berger SL, Zhu H. Protein acetylation microarray reveals that NuA4 controls key metabolic target regulating gluconeogenesis. *Cell.* 2009;136(6):1073–1084.
61. He W, Newman JC, Wang MZ, Ho L, Verdin E. Mitochondrial sirtuins: regulators of protein acylation and metabolism. *Trends Endocrinol Metab.* 2012;23(9):467–476.
62. Cosgrove MS, Bever K, Avalos JL, Muhammad S, Zhang X, Wolberger C. The structural basis of sirtuin substrate affinity. *Biochemistry.* 2006;45(24):7511–7521.
63. Rauh D, Fischer F, Gertz M, Lakshminarasimhan M, Bergbrede T, Aladini F, Kambach C, Becker CF, Zerweck J, Schutkowski M, Steegborn C. An acetylome peptide microarray reveals specificities and deacetylation substrates for all human sirtuin isoforms. *Nat Commun.* 2013;4:2327.
64. Blander G, Olejnik J, Krzymanska-Olejnik E, McDonagh T, Haigis M, Yaffe MB, Guarente L. SIRT1 shows no substrate specificity in vitro. *J Biol Chem.* 2005;280 (11):9780–9785.
65. Garske AL, Denu JM. SIRT1 top 40 hits: use of one-bead, one-compound acetyl-peptide libraries and quantum dots to probe deacetylase specificity. *Biochemistry.* 2006;45(1):94–101.
66. AbouElfetouh A, Kuhn ML, Hu LI, Scholle MD, Sorensen DJ, Sahu AK, Becher D, Antelmann H, Mrksich M, Anderson WF, Gibson BW, Schilling B, Wolfe AJ. The *E. coli* sirtuin CobB shows no preference for enzymatic and nonenzymatic lysine acetylation substrate sites. *MicrobiologyOpen.* 2015;4(1):66–83.
67. Smith BC, Settles B, Hallows WC, Craven MW, Denu JM. SIRT3 substrate specificity determined by peptide arrays and machine learning. *ACS Chem Biol.* 2010;6 (2):146–157.
68. Haigis MC, Sinclair DA. Mammalian sirtuins: biological insights and disease relevance. *Annu Rev Pathol.* 2010;5:253–295.
69. Giblin W, Skinner ME, Lombard DB. Sirtuins: guardians of mammalian healthspan. *Trends Genet.* 2014;30(7):271–286.
70. Schiedel M, Robaa D, Rumpf T, Sippl W, Jung M. The current state of NAD^+-dependent histone deacetylases (sirtuins) as novel therapeutic targets. *Med Res Rev.* 2017. 10.1002/med.21436.
71. Howitz KT, Bitterman KJ, Cohen HY, Lamming DW, Lavu S, Wood JG, Zipkin RE, Chung P, Kisielewski A, Zhang LL, Scherer B, Sinclair DA. Small molecule activators of sirtuins extend *Saccharomyces cerevisiae* lifespan. *Nature.* 2003;425(6954):191–196.
72. Wood JG, Rogina B, Lavu S, Howitz K, Helfand SL, Tatar M, Sinclair D. Sirtuin activators mimic caloric restriction and delay ageing in metazoans. *Nature.* 2004;430 (7000):686–689.
73. Milne JC, Lambert PD, Schenk S, Carney DP, Smith JJ, Gagne DJ, Jin L, Boss O, Perni RB, Vu CB, Bemis JE, Xie R, Disch JS, Ng PY, Nunes JJ, Lynch AV, Yang H, Galonek H, Israelian K, Choy W, Iffland A, Lavu S, Medvedik O, Sinclair DA, Olefsky JM, Jirousek MR, Elliott PJ, Westphal CH. Small molecule activators of SIRT1 as therapeutics for the treatment of type 2 diabetes. *Nature.* 2007;450(7170):712–716.
74. Kaeberlein M, McVey M, Guarente L. The *SIR2/3/4* complex and *SIR2* alone promote longevity in *Saccharomyces cerevisiae* by two different mechanisms. *Genes Dev.* 1999;13(19):2570–2580.

75. Tissenbaum HA, Guarente L. Increased dosage of a *sir-2* gene extends lifespan in *Caenorhabditis elegans*. *Nature*. 2001;410(6825):227–230.
76. Outeiro TF, Kontopoulos E, Altmann SM, Kufareva I, Strathearn KE, Amore AM, Volk CB, Maxwell MM, Rochet JC, McLean PJ, Young AB, Abagyan R, Feany MB, Hyman BT, Kazantsev AG. Sirtuin 2 inhibitors rescue α-synuclein-mediated toxicity in models of Parkinson's disease. *Science*. 2007;317(5837):516–519.
77. Vaziri H, Dessain SK, Eaton EN, Imai S-I, Frye RA, Pandita TK, Guarente L, Weinberg RA. $hSIR2^{SIRT1}$ functions as an NAD-dependent p53 deacetylase. *Cell*. 2001;107(2): 149–159.
78. Heltweg B, Gatbonton T, Schuler AD, Posakony J, Li H, Goehle S, Kollipara R, Depinho RA, Gu Y, Simon JA, Bedalov A. Antitumor activity of a small-molecule inhibitor of human silent information regulator 2 enzymes. *Cancer Res*. 2006;66 (8):4368–4377.
79. Ashraf N, Zino S, Macintyre A, Kingsmore D, Payne AP, George WD, Shiels PG. Altered sirtuin expression is associated with node-positive breast cancer. *Br J Cancer*. 2006;95(8):1056–1061.
80. Süssmuth SD, Haider S, Landwehrmeyer GB, Farmer R, Frost C, Tripepi G, Andersen CA, Di Bacco M, Lamanna C, Diodato E, Massai L, Diamanti D, Mori E, Magnoni L, Dreyhaupt J, Schiefele K, Craufurd D, Saft C, Rudzinska M, Ryglewicz D, Orth M, Brzozy S, Baran A, Pollio G, Andre R, Tabrizi SJ, Darpo B, Westerberg G. Paddington Consortium.An exploratory double-blind, randomized clinical trial with selisistat, a SirT1 inhibitor, in patients with Huntington's disease. *Br J Clin Pharmacol*. 2015;79 (3):465–476.
81. Bonkowski MS, Sinclair DA. Slowing ageing by design: the rise of NAD^+ and sirtuin-activating compounds. *Nat Rev Mol Cell Biol*. 2016;17(11):679–690.
82. Disch JS, Evindar G, Chiu CH, Blum CA, Dai H, Jin L, Schuman E, Lind KE, Belyanskaya SL, Deng J, Coppo F, Aquilani L, Graybill TL, Cuozzo JW, Lavu S, Mao C, Vlasuk GP, Perni RB. Discovery of thieno[3,2-*d*]pyrimidine-6-carboxamides as potent inhibitors of SIRT1, SIRT2, and SIRT3. *J Med Chem*. 2013;56(9):3666–3679.
83. Napper AD, Hixon J, McDonagh T, Keavey K, Pons JF, Barker J, Yau WT, Amouzegh P, Flegg A, Hamelin E, Thomas RJ, Kates M, Jones S, Navia MA, Saunders JO, DiStefano PS, Curtis R. Discovery of indoles as potent and selective inhibitors of the deacetylase SIRT1. *J Med Chem*. 2005;48(25):8045–8054.
84. Schlicker C, Boanca G, Lakshminarasimhan M, Steegborn C. Structure-based development of novel sirtuin inhibitors. *Aging*. 2011;3(9):852–872.
85. Pulla VK, Alvala M, Sriram DS, Viswanadha S, Sriram D, Yogeeswari P. Structure-based drug design of small molecule SIRT1 modulators to treat cancer and metabolic disorders. *J Mol Graphics Modell*. 2014;52:46–56.
86. Sacconnay L, Ryckewaert L, Randazzo GM, Petit C, Passos Cdos S, Jachno J, Michailovienė V, Zubrienė A, Matulis D, Carrupt PA, Simões-Pires CA, Nurisso A. 5-Benzylidene-hydantoin is a new scaffold for SIRT inhibition: from virtual screening to activity assays. *Eur J Pharm Sci*. 2016;85:59–67.
87. Uciechowska U, Schemies J, Neugebauer RC, Huda EM, Schmitt ML, Meier R, Verdin E, Jung M, Sippl W. Thiobarbiturates as sirtuin inhibitors: virtual screening, free-energy calculations, and biological testing. *ChemMedChem*. 2008;3(12):1965–1976.
88. Jiang Y, Liu J, Chen D, Yan L, Zheng W. Sirtuin inhibition: strategies, inhibitors, and therapeutic potential. *Trends Pharmacol Sci*. 2017;38(5):459–472.
89. Chen B, Zang W, Wang J, Huang Y, He Y, Yan L, Liu J, Zheng W. The chemical biology of sirtuins. *Chem Soc Rev*. 2015;44(15):5246–5264.
90. Schutkowski M, Fischer F, Roessler C, Steegborn C. New assays and approaches for discovery and design of sirtuin modulators. *Expert Opin Drug Discov*. 2014;9(2): 183–199.

91. Schemies J, Uciechowska U, Sippl W, Jung M. NAD^+-dependent histone deacetylases (sirtuins) as novel therapeutic targets. *Med Res Rev.* 2010;30(6):861–889.
92. Ng PY, Bemis JE, Disch JS, Vu CB, Oalmann Cj. Lynch AV, Carney DP, Riera TV, Song J, Smith JJ, Lavu S, Tornblom A, Duncan M, Yeager M, Kriksciukaite K, Gupta A, Suri V, Elliot PJ, Milne JC, Nunes JJ, Jirousek MR, Vlasuk GP, Ellis JL, Perni RB. The identification of the SIRT1 activator SRT2104 as a clinical candidate. *Lett Drug Des Discov.* 2013;10(9):793–797.
93. Nayagam VM, Wang X, Tan YC, Poulsen A, Goh KC, Ng T, Wang H, Song HY, Ni B, Entzeroth M, Stünkel W. SIRT1 modulating compounds from high-throughput screening as anti-inflammatory and insulin-sensitizing agents. *J Biomol Screen.* 2006;11 (8):959–967.
94. Rumpf T, Schiedel M, Karaman B, Roessler C, North BJ, Lehotzky A, Oláh J, Ladwein KI, Schmidtkunz K, Gajer M, Pannek M, Steegborn C, Sinclair DA, Gerhardt S, Ovádi J, Schutkowski M, Sippl W, Einsle O, Jung M. Selective Sirt2 inhibition by ligand-induced rearrangement of the active site. *Nat Commun.* 2015;6:6263.
95. Smith BC, Denu JM. Mechanism-based inhibition of Sir2 deacetylases by thioacetyl-lysine peptide. *Biochemistry.* 2007;46(50):14478–14486.
96. Smith BC, Denu JM. Acetyl-lysine analog peptides as mechanistic probes of protein deacetylases. *J Biol Chem.* 2007;282(51):37256–37265.
97. Smith BC, Denu JM. Sir2 deacetylases exhibit nucleophilic participation of acetyl-lysine in NAD^+ cleavage. *J Am Chem Soc.* 2007;129(18):5802–5803.
98. Zhou Y, Zhang H, He B, Du J, Lin H, Cerione RA, Hao Q. The bicyclic intermediate structure provides insights into the desuccinylation mechanism of human sirtuin 5 (SIRT5). *J Biol Chem.* 2012;287(34):28307–28314.
99. Wang Y, Fung YM, Zhang W, He B, Chung MW, Jin J, Hu J, Lin H, Hao Q. Deacylation mechanism by SIRT2 revealed in the 1′-SH-2′-O-myristoyl intermediate structure. *Cell Chem Biol.* 2017;24(3):339–345.
100. Avalos JL, Celic I, Muhammad S, Cosgrove MS, Boeke JD, Wolberger C. Structure of a Sir2 enzyme bound to an acetylated p53 peptide. *Mol Cell.* 2002;10(3):523–535.
101. Avalos JL, Boeke JD, Wolberger C. Structural basis for the mechanism and regulation of Sir2 enzymes. *Mol Cell.* 2004;13(5):639–648.
102. Borra MT, Langer MR, Slama JT, Denu JM. Substrate specificity and kinetic mechanism of the Sir2 family of NAD^+-dependent histone/protein deacetylases. *Biochemistry.* 2004;43(30):9877–9887.
103. Gertz M, Fischer F, Nguyen GT, Lakshminarasimhan M, Schutkowski M, Weyand M, Steegborn C. Ex-527 inhibits sirtuins by exploiting their unique NAD^+-dependent deacetylation mechanism. *Proc Natl Acad Sci USA.* 2013;110(30): E2772–E2781.
104. Smith BC, Denu JM. Sir2 protein deacetylases: evidence for chemical intermediates and functions of a conserved histidine. *Biochemistry.* 2006;45(1):272–282.
105. Suzuki T, Asaba T, Imai E, Tsumoto H, Nakagawa H, Miyata N. Identification of a cell-active non-peptide sirtuin inhibitor containing *N*-thioacetyl lysine. *Bioorg Med Chem Lett.* 2009;19(19):5670–5672.
106. Hawse WF, Hoff KG, Fatkins DG, Daines A, Zubkova OV, Schramm VL, Zheng W, Wolberger C. Structural insights into intermediate steps in the Sir2 deacetylation reaction. *Structure.* 2008;16(9):1368–1377.
107. Zhao K, Harshaw R, Chai X, Marmorstein R. Structural basis for nicotinamide cleavage and ADP-ribose transfer by NAD^+-dependent Sir2 histone/protein deacetylases. *Proc Natl Acad Sci USA.* 2004;101(23):8563–8568.
108. Jackson MD, Schmidt MT, Oppenheimer NJ, Denu JM. Mechanism of nicotinamide inhibition and transglycosidation by Sir2 histone/protein deacetylases. *J Biol Chem.* 2003;278(51):50985–50998.

109. Hoff KG, Avalos JL, Sens K, Wolberger C. Insights into the sirtuin mechanism from ternary complexes containing NAD^+ and acetylated peptide. *Structure.* 2006;14(8): 1231–1240.
110. Sauve AA. Sirtuin chemical mechanisms. *Biochim Biophys Acta.* 2010;1804 (8):1591–1603.
111. Min J, Landry J, Sternglanz R, Xu RM. Crystal structure of a SIR2 homolog-NAD complex. *Cell.* 2001;105(2):269–279.
112. Fatkins DG, Monnot AD, Zheng W. N^ε-thioacetyl-lysine: a multi-facet functional probe for enzymatic protein lysine N^ε-deacetylation. *Bioorg Med Chem Lett.* 2006;16 (14):3651–3656.
113. Hirsch BM, Hao Y, Li X, Wesdemiotis C, Wang Z, Zheng W. A mechanism-based potent sirtuin inhibitor containing N^ε-thiocarbamoyl-lysine (TuAcK). *Bioorg Med Chem Lett.* 2011;21(16):4753–4757.
114. Lin H. Cornell University, assignee; Thiourea compounds and their use as inhibitors of SIRT2 or SIRT5. International Publication Number WO 2014/197775 A1. December 11, 2014
115. Zang W, Hao Y, Wang Z, Zheng W. Novel thiourea-based sirtuin inhibitory warheads. *Bioorg Med Chem Lett.* 2015;25(16):3319–3324.
116. Liu J, Zheng W. Cyclic peptide-based potent human SIRT6 inhibitors. *Org Biomol Chem.* 2016;14(25):5928–5935.
117. Liu J, Huang Y, Zheng W. A selective cyclic peptidic human SIRT5 inhibitor. *Molecules.* 2016;21(9):1217.
118. Morimoto J, Hayashi Y, Suga H. Discovery of macrocyclic peptides armed with a mechanism-based warhead: isoform-selective inhibition of human deacetylase SIRT2. *Angew Chem Int Ed.* 2012;51(14):3423–3427.
119. Hirsch BM, Du Z, Li X, Sylvester JA, Wesdemiotis C, Wang Z, Zheng W. Potent sirtuin inhibition bestowed by L-2-amino-7-carboxamidoheptanoic acid (L-ACAH), a N^ε-acetyl-lysine analog. *Med Chem Comm.* 2011;2(4):291–299.
120. He Y, Yan L, Zang W, Zheng W. Novel sirtuin inhibitory warheads derived from the N^ε-acetyl-lysine analog L-2-amino-7-carboxamidoheptanoic acid. *Org Biomol Chem.* 2015;13(42):10442–10450.
121. Huhtiniemi T, Suuronen T, Lahtela-Kakkonen M, Bruijn T, Jääskeläinen S, Poso A, Salminen A, Leppänen J, Jarho E. N^ε-modified lysine containing inhibitors for SIRT1 and SIRT2. *Bioorg Med Chem.* 2010;18(15):5616–5625.
122. Kokkonen P, Rahnasto-Rilla M, Kiviranta PH, Huhtiniemi T, Laitinen T, Poso A, Jarho E, Lahtela-Kakkonen M. Peptides and pseudopeptides as SIRT6 deacetylation inhibitors. *ACS Med Chem Lett.* 2012;3(12):969–974.
123. Cupp-Sutton KA, Ashby MT. Biological chemistry of hydrogen selenide. *Antioxidants.* 2016;5(4):42.
124. Dancy BC, Ming SA, Papazyan R, Jelinek CA, Majumdar A, Sun Y, Dancy BM, Drury 3rd WJ, Cotter RJ, Taverna SD, Cole PA. Azalysine analogues as probes for protein lysine deacetylation and demethylation. *J Am Chem Soc.* 2012;134(11):5138–5148.
125. Madsen AS, Andersen C, Daoud M, Anderson KA, Laursen JS, Chakladar S, Huynh FK, Colaço AR, Backos DS, Fristrup P, Hirschey MD, Olsen CA. Investigating the sensitivity of NAD+-dependent sirtuin deacylation activities to NADH. *J Biol Chem.* 2016;291(13):7128–7141.
126. Feldman JL, Baeza J, Denu JM. Activation of the protein deacetylase SIRT6 by long-chain fatty acids and widespread deacylation by mammalian sirtuins. *J Biol Chem.* 2013;288(43):31350–31356.
127. Galleano I, Schiedel M, Jung M, Madsen AS, Olsen CA. A continuous, fluorogenic sirtuin 2 deacylase assay: substrate screening and inhibitor evaluation. *J Med Chem.* 2016;59(3):1021–1031.

128. Madsen AS, Olsen CA. Substrates for efficient fluorometric screening employing the NAD-dependent sirtuin 5 lysine deacylase (KDAC) enzyme. *J Med Chem.* 2012;55 (11):5582–5590.
129. Tong Z, Wang M, Wang Y, Kim DD, Grenier JK, Cao J, Sadhukhan S, Hao Q, Lin H. SIRT7 is an RNA-activated protein lysine deacylase. *ACS Chem Biol.* 2016;12(1):300–310.
130. Tong Z, Wang Y, Zhang X, Kim DD, Sadhukhan S, Hao Q, Lin H. SIRT7 is activated by DNA and deacetylates histone H3 in the chromatin context. *ACS Chem Biol.* 2016;11(3):742–747.
131. Li L, Shi L, Yang S, Yan R, Zhang D, Yang J, He L, Li W, Yi X, Sun L, Liang J, Cheng Z, Shi L, Shang Y, Yu W. SIRT7 is a histone desuccinylase that functionally links to chromatin compaction and genome stability. *Nat Commun.* 2016;7:12235.
132. Chalkiadaki A, Guarente L. The multifaceted functions of sirtuins in cancer. *Nat Rev Cancer.* 2015;15(10):608–624.
133. Allfrey V, Faulkner R, Mirsky A. Acetylation and methylation of histones and their possible role in the regulation of RNA synthesis. *Proc Natl Acad Sci USA.* 1964;51(5): 786–794.
134. Tse C, Sera T, Wolffe AP, Hansen JC. Disruption of higher-order folding by core histone acetylation dramatically enhances transcription of nucleosomal arrays by RNA polymerase III. *Mol Cell Biol.* 1998;18(8):4629–4638.
135. Shogren-Knaak M, Ishii H, Sun J-M, Pazin MJ, Davie JR, Peterson CL. Histone H4-K16 acetylation controls chromatin structure and protein interactions. *Science.* 2006;311(5762):844–847.
136. Xu H, Chen XY, Xu XL, Shi RY, Suo SS, Cheng KY, Zheng ZG, Wang MX, Wang LY, Zhao Y, Tian B, Hua YJ. Lysine acetylation and succinylation in HeLa cells and their essential roles in response to UV-induced stress. *Sci Rep.* 2016;6:30212.
137. Chen Y, Zhao W, Yang JS, Cheng Z, Luo H, Lu Z, Tan M, Gu W, Zhao Y. Quantitative acetylome analysis reveals the roles of SIRT1 in regulating diverse substrates and cellular pathways. *Mol Cell Proteomics.* 2012;11(10):1048–1062.
138. Lin S-J, Defossez P-A, Guarente L. Requirement of NAD and SIR2 for life-span extension by calorie restriction in *Saccharomyces cerevisiae. Science.* 2000;289(5487): 2126–2128.
139. Langley E, Pearson M, Faretta M, Bauer UM, Frye RA, Minucci S, Pelicci PG, Kouzarides T. Human SIR2 deacetylates p53 and antagonizes PML/p53-induced cellular senescence. *EMBO J.* 2002;21(10):2383–2396.
140. Zhang F, Wang S, Gan L, Vosler PS, Gao Y, Zigmond MJ, Chen J. Protective effects and mechanisms of sirtuins in the nervous system. *Prog Neurobiol.* 2011;95(3):373–395.
141. Luo J, Nikolaev AY, Imai S-i. Chen D, Su F, Shiloh A, Guarente L, Gu W. Negative control of p53 by Sir2α promotes cell survival under stress. *Cell.* 2001;107(2):137–148.
142. Duffy MJ, Synnott NC, McGowan PM, Crown J, O'Connor D, Gallagher WM. p53 as a target for the treatment of cancer. *Cancer Treat Rev.* 2014;40(10):1153–1160.
143. Li L, Wang L, Li L, Wang Z, Ho Y, McDonald T, Holyoake TL, Chen W, Bhatia R. Activation of p53 by SIRT1 inhibition enhances elimination of CML leukemia stem cells in combination with imatinib. *Cancer Cell.* 2012;21(2):266–281.
144. Wu X, Cao N, Fenech M, Wang X. Role of sirtuins in maintenance of genomic stability: relevance to cancer and healthy aging. *DNA Cell Biol.* 2016;35(10):542–575.
145. Brodeur GM. Neuroblastoma: biological insights into a clinical enigma. *Nat Rev Cancer.* 2003;3(3):203–216.
146. Marshall GM, Liu PY, Gherardi S, Scarlett CJ, Bedalov A, Xu N, Iraci N, Valli E, Ling D, Thomas W, van Bekkum M, Sekyere E, Jankowski K, Trahair T, MacKenzie KL, Haber M, Norris MD, Biankin AV, Perini G, Liu T. SIRT1 promotes N-Myc oncogenesis through a positive feedback loop involving the effects of MKP3 and ERK on N-Myc protein stability. *PLoS Genet.* 2011;7(6):e1002135.

147. Munro S, Carr SM, La Thangue NB. Diversity within the pRb pathway: is there a code of conduct? *Oncogene*. 2012;31(40):4343–4352.
148. Wong S, Weber JD. Deacetylation of the retinoblastoma tumour suppressor protein by SIRT1. *Biochem J*. 2007;407(3):451–460.
149. de Oliveira RM, Sarkander J, Kazantsev AG, Outeiro TF. SIRT2 as a therapeutic target for age-related disorders. *Front Pharmacol*. 2012;3:82.
150. Li Y, Matsumori H, Nakayama Y, Osaki M, Kojima H, Kurimasa A, Ito H, Mori S, Katoh M, Oshimura M. SIRT2 down-regulation in HeLa can induce p53 accumulation via p38 MAPK activation-dependent p300 decrease, eventually leading to apoptosis. *Genes Cells*. 2011;16(1):34–45.
151. Hua X, Yokoyama C, Wu J, Briggs MR, Brown MS, Goldstein JL, Wang X. SREBP-2, a second basic-helix-loop-helix-leucine zipper protein that stimulates transcription by binding to a sterol regulatory element. *Proc Natl Acad Sci USA*. 1993;90(24):11603–11607.
152. Luthi-Carter R, Taylor DM, Pallos J, Lambert E, Amore A, Parker A, Moffitt H, Smith DL, Runne H, Gokce O, Kuhn A, Xiang Z, Maxwell MM, Reeves SA, Bates GP, Neri C, Thompson LM, Marsh JL, Kazantsev AG. SIRT2 inhibition achieves neuroprotection by decreasing sterol biosynthesis. *Proc Natl Acad Sci USA*. 2010;107(17):7927–7932.
153. Lombard DB, Alt FW, Cheng HL, Bunkenborg J, Streeper RS, Mostoslavsky R, Kim J, Yancopoulos G, Valenzuela D, Murphy A, Yang Y, Chen Y, Hirschey MD, Bronson RT, Haigis M, Guarente LP, Farese Jr RV, Weissman S, Verdin E, Schwer B. Mammalian Sir2 homolog SIRT3 regulates global mitochondrial lysine acetylation. *Mol Cell Biol*. 2007;27(24):8807–8814.
154. Alhazzazi TY, Kamarajan P, Joo N, Huang JY, Verdin E, D'Silva NJ, Kapila YL. Sirtuin-3 (SIRT3), a novel potential therapeutic target for oral cancer. *Cancer*. 2011;117(8): 1670–1678.
155. Sundaresan NR, Samant SA, Pillai VB, Rajamohan SB, Gupta MP. SIRT3 is a stress-responsive deacetylase in cardiomyocytes that protects cells from stress-mediated cell death by deacetylation of Ku70. *Mol Cell Biol*. 2008;28(20):6384–6401.
156. Cohen HY, Lavu S, Bitterman KJ, Hekking B, Imahiyerobo TA, Miller C, Frye R, Ploegh H, Kessler BM, Sinclair DA. Acetylation of the C terminus of Ku70 by CBP and PCAF controls Bax-mediated apoptosis. *Mol Cell*. 2004;13(5):627–638.
157. Alhazzazi TY, Kamarajan P, Verdin E, Kapila YL. SIRT3 and cancer: tumor promoter or suppressor? *Biochim Biophys Acta Rev Cancer*. 2011;1816(1):80–88.
158. Finley LW, Haigis MC. Metabolic regulation by SIRT3: implications for tumorigenesis. *Trends Mol Med*. 2012;18(9):516–523.
159. Gertz M, Steegborn C. Using mitochondrial sirtuins as drug targets: disease implications and available compounds. *Cell Mol Life Sci*. 2016;73(15):2871–2896.
160. Liszt G, Ford E, Kurtev M, Guarente L. Mouse Sir2 homolog SIRT6 is a nuclear ADP-ribosyltransferase. *J Biol Chem*. 2005;280(22):21313–21320.
161. Michishita E, McCord RA, Berber E, Kioi M, Padilla-Nash H, Damian M, Cheung P, Kusumoto R, Kawahara TLA, Barrett JC, Chang HY, Bohr VA, Ried T, Gozani O, Chua KF. SIRT6 is a histone H3 lysine 9 deacetylase that modulates telomeric chromatin. *Nature*. 2008;452(7186):492–496.
162. Zhong L, D'Urso A, Toiber D, Sebastian C, Henry RE, Vadysirisack DD, Guimaraes A, Marinelli B, Wikstrom JD, Nir T, Clish CB, Vaitheesvaran B, Iliopoulos O, Kurland I, Dor Y, Weissleder R, Shirihai OS, Ellisen LW, Espinosa JM, Mostoslavsky R. The histone deacetylase Sirt6 regulates glucose homeostasis via Hif1α. *Cell*. 2010;140(2):280–293.
163. Kaidi A, Weinert BT, Choudhary C, Jackson SP. Human SIRT6 promotes DNA-end resection through CtIP deacetylation. *Science*. 2010;329(5997):1348–1353.
164. Mao Z, Hine C, Tian X, Van Meter M, Au M, Vaidya A, Seluanov A, Gorbunova V. SIRT6 promotes DNA repair under stress by activating PARP1. *Science*. 2011;332 (6036):1443–1446.

165. Tasselli L, Zheng W, Chua KF. SIRT6: novel mechanisms and links to aging and disease. *Trends Endocrinol Metab.* 2017;28(3):168–185.
166. Kim E-J, Juhnn Y-S. Cyclic AMP signaling reduces sirtuin 6 expression in non-small cell lung cancer cells by promoting ubiquitin-proteasomal degradation via inhibition of the Raf-MEK-ERK (Raf/mitogen-activated extracellular signal-regulated kinase/extracellular signal-regulated kinase) pathway. *J Biol Chem.* 2015;290(15):9604–9613.
167. Liu Y, Xie QR, Wang B, Shao J, Zhang T, Liu T, Huang G, Xia W. Inhibition of SIRT6 in prostate cancer reduces cell viability and increases sensitivity to chemotherapeutics. *Protein Cell.* 2013;4(9):702–710.
168. Ming M, Han W, Zhao B, Sundaresan NR, Deng C-X, Gupta MP, He Y-Y. SIRT6 promotes COX-2 expression and acts as an oncogene in skin cancer. *Cancer Res.* 2014;74(20):5925–5933.
169. Ran LK, Chen Y, Zhang ZZ, Tao NN, Ren JH, Zhou L, Tang H, Chen X, Chen K, Li WY, Huang AL, Chen J. SIRT6 overexpression potentiates apoptosis evasion in hepatocellular carcinoma via BCL2-associated X protein-dependent apoptotic pathway. *Clin Cancer Res.* 2016;22(13):3372–3382.
170. Ford E, Voit R, Liszt G, Magin C, Grummt I, Guarente L. Mammalian Sir2 homolog SIRT7 is an activator of RNA polymerase I transcription. *Genes Dev.* 2006;20 (9):1075–1080.
171. Chen S, Seiler J, Santiago-Reichelt M, Felbel K, Grummt I, Voit R. Repression of RNA polymerase I upon stress is caused by inhibition of RNA-dependent deacetylation of PAF53 by SIRT7. *Mol Cell.* 2013;52(3):303–313.
172. Chen S, Blank MF, Iyer A, Huang B, Wang L, Grummt I, Voit R. SIRT7-dependent deacetylation of the U3-55k protein controls pre-rRNA processing. *Nat Commun.* 2016;7:10734.
173. Yu H, Ye W, Wu J, Meng X, Liu R-y. Ying X, Zhou Y, Wang H, Pan C, Huang W. Overexpression of Sirt7 exhibits oncogenic property and serves as a prognostic factor in colorectal cancer. *Clin Cancer Res.* 2014;20(13):3434–3445.
174. Blank MF, Grummt I. The seven faces of SIRT7. *Transcription.* 2017;8(2):67–74.
175. Barber MF, Michishita-Kioi E, Xi Y, Tasselli L, Kioi M, Moqtaderi Z, Tennen RI, Paredes S, Young NL, Chen K, Struhl K, Garcia BA, Gozani O, Li W, Chua KF. SIRT7 links H3K18 deacetylation to maintenance of oncogenic transformation. *Nature.* 2012;487(7405):114–118.
176. Fatkins DG, Zheng W. Substituting N^{ϵ}-thioacetyl-lysine for N^{ϵ}-acetyl-lysine in peptide substrates as a general approach to inhibiting human NAD^+-dependent protein deacetylases. *Int J Mol Sci.* 2008;9(1):1–11.
177. Asaba T, Suzuki T, Ueda R, Tsumoto H, Nakagawa H, Miyata N. Inhibition of human sirtuins by in situ generation of an acetylated lysine-ADP-ribose conjugate. *J Am Chem Soc.* 2009;131(20):6989–6996.
178. Kiviranta PH, Suuronen T, Wallén EA, Leppänen J, Tervonen J, Kyrylenko S, Salminen A, Poso A, Jarho EM. N^{ϵ}-thioacetyl-lysine-containing tri-, tetra-, and pentapeptides as SIRT1 and SIRT2 inhibitors. *J Med Chem.* 2009;52(7):2153–2156.
179. Hirsch BM, Gallo CA, Du Z, Wang Z, Zheng W. Discovery of potent, proteolytically stable, and cell permeable human sirtuin peptidomimetic inhibitors containing N^{ϵ}-thioacetyl-lysine. *Med Chem Comm.* 2010;1(3):233–238.
180. Mellini P, Kokkola T, Suuronen T, Salo HS, Tolvanen L, Mai A, Lahtela-Kakkonen M, Jarho EM. Screen of pseudopeptidic inhibitors of human sirtuins 1-3: two lead compounds with antiproliferative effects in cancer cells. *J Med Chem.* 2013;56(17):6681–6695.
181. Huhtiniemi T, Salo HS, Suuronen T, Poso A, Salminen A, Leppänen J, Jarho E, Lahtela-Kakkonen M. Structure-based design of pseudopeptidic inhibitors for SIRT1 and SIRT2. *J Med Chem.* 2011;54(19):6456–6468.

182. Jing H, Hu J, He B, Negrón Abril YL, Stupinski J, Weiser K, Carbonaro M, Chiang YL, Southard T, Giannakakou P, Weiss RS, Lin H. A SIRT2-selective inhibitor promotes c-Myc oncoprotein degradation and exhibits broad anticancer activity. *Cancer Cell.* 2016;29(3):297–310.
183. Craik DJ, Fairlie DP, Liras S, Price D. The future of peptide-based drugs. *Chem Biol Drug Des.* 2013;81(1):136–147.
184. Chen B, Wang J, Huang Y, Zheng W. Human SIRT3 tripeptidic inhibitors containing N^{ε}-thioacetyl-lysine. *Bioorg Med Chem Lett.* 2015;25(17):3481–3487.
185. Huang Y, Liu J, Yan L, Zheng W. Simple N^{ε}-thioacetyl-lysine-containing cyclic peptides exhibiting highly potent sirtuin inhibition. *Bioorg Med Chem Lett.* 2016;26(6):1612–1617.
186. Wang J, Zang W, Liu J, Zheng W. Bivalent SIRT1 inhibitors. *Bioorg Med Chem Lett.* 2017;27(2):180–186.
187. Alhazzazi TY, Kamarajan P, Xu Y, Ai T, Chen L, Verdin E, Kapila YL. A novel sirtuin-3 inhibitor, LC-0296, inhibits cell survival and proliferation, and promotes apoptosis of head and neck cancer cells. *Anticancer Res.* 2016;36(1):49–60.
188. Liu B, Lin Y, Darwanto A, Song X, Xu G, Zhang K. Identification and characterization of propionylation at histone H3 lysine 23 in mammalian cells. *J Biol Chem.* 2009;284(47):32288–32295.
189. Goudarzi A, Zhang D, Huang H, Barral S, Kwon OK, Qi S, Tang Z, Buchou T, Vitte AL, He T, Cheng Z, Montellier E, Gaucher J, Curtet S, Debernardi A, Charbonnier G, Puthier D, Petosa C, Panne D, Rousseaux S, Roeder RG, Zhao Y, Khochbin S. Dynamic competing histone H4 K5K8 acetylation and butyrylation are hallmarks of highly active gene promoters. *Mol Cell.* 2016;62(2):169–180.
190. Garrity J, Gardner JG, Hawse W, Wolberger C, Escalante-Semerena JC. *N*-lysine propionylation controls the activity of propionyl-CoA synthetase. *J Biol Chem.* 2007;282(41):30239–30245.
191. Madsen AS, Olsen CA. Profiling of substrates for zinc-dependent lysine deacylase enzymes: HDAC3 exhibits decrotonylase activity in vitro. *Angew Chem Int Ed.* 2012;51(36):9083–9087.
192. Hedo JA, Collier E, Watkinson A. Myristyl and palmityl acylation of the insulin receptor. *J Biol Chem.* 1987;262(3):954–957.
193. Pillai S, Baltimore D. Myristoylation and the post-translational acquisition of hydrophobicity by the membrane immunoglobulin heavy-chain polypeptide in B lymphocytes. *Proc Natl Acad Sci USA.* 1987;84(21):7654–7658.
194. Van Gool F, Galli M, Gueydan C, Kruys V, Prevot PP, Bedalov A, Mostoslavsky R, Alt FW, De Smedt T, Leo O. Intracellular NAD levels regulate tumor necrosis factor protein synthesis in a sirtuin-dependent manner. *Nat Med.* 2009;15(2):206–210.
195. Liu Z, Yang T, Li X, Peng T, Hang HC, Li XD. Integrative chemical biology approaches for identification and characterization of "erasers" for fatty-acid-acylated lysine residues within proteins. *Angew Chem Int Ed.* 2015;54(4):1149–1152.
196. Teng YB, Jing H, Aramsangtienchai P, He B, Khan S, Hu J, Lin H, Hao Q. Efficient demyristoylase activity of SIRT2 revealed by kinetic and structural studies. *Sci Rep.* 2015;5:8529.
197. He B, Hu J, Zhang X, Lin H. Thiomyristoyl peptides as cell-permeable Sirt6 inhibitors. *Org Biomol Chem.* 2014;12(38):7498–7502.
198. Rardin MJ, He W, Nishida Y, Newman JC, Carrico C, Danielson SR, Guo A, Gut P, Sahu AK, Li B, Uppala R, Fitch M, Riiff T, Zhu L, Zhou J, Mulhern D, Stevens RD, Ilkayeva OR, Newgard CB, Jacobson MP, Hellerstein M, Goetzman ES, Gibson BW, Verdin E. SIRT5 regulates the mitochondrial lysine succinylome and metabolic networks. *Cell Metab.* 2013;18(6):920–933.

199. Zhang Y, Bharathi SS, Rardin MJ, Lu J, Maringer KV, Sims-Lucas S, Prochownik EV, Gibson BW, Goetzman ES. Lysine desuccinylase SIRT5 binds to cardiolipin and regulates the electron transport chain. *J Biol Chem.* 2017;292(24): 10239–10249.
200. Park J, Chen Y, Tishkoff DX, Peng C, Tan M, Dai L, Xie Z, Zhang Y, Zwaans BMM, Skinner ME, Lombard DB, Zhao Y. SIRT5-mediated lysine desuccinylation impacts diverse metabolic pathways. *Mol Cell.* 2013;50(6):919–930.
201. Nakagawa T, Lomb DJ, Haigis MC, Guarente L. SIRT5 deacetylates carbamoyl phosphate synthetase 1 and regulates the urea cycle. *Cell.* 2009;137(3):560–570.
202. Nishida Y, Rardin MJ, Carrico C, He W, Sahu AK, Gut P, Najjar R, Fitch M, Hellerstein M, Gibson BW, Verdin E. SIRT5 regulates both cytosolic and mitochondrial protein malonylation with glycolysis as a major target. *Mol Cell.* 2015;59(2): 321–332.
203. Lin ZF, Xu HB, Wang JY, Lin Q, Ruan Z, Liu FB, Jin W, Huang HH, Chen X. SIRT5 desuccinylates and activates SOD1 to eliminate ROS. *Biochem Biophys Res Commun.* 2013;441(1):191–195.
204. Glasauer A, Sena LA, Diebold LP, Mazar AP, Chandel NS. Targeting SOD1 reduces experimental non-small-cell lung cancer. *J Clin Invest.* 2014;124(1):117–128.
205. Lu W, Zuo Y, Feng Y, Zhang M. SIRT5 facilitates cancer cell growth and drug resistance in non-small cell lung cancer. *Tumour Biol.* 2014;35(11):10699–10705.
206. Xiangyun Y, Xiaomin N, Linping G, Yunhua X, Ziming L, Yongfeng Y, Zhiwei C, Shun L. Desuccinylation of pyruvate kinase M2 by SIRT5 contributes to antioxidant response and tumor growth. *Oncotarget.* 2017;8(4):6984–6993.
207. Roessler C, Nowak T, Pannek M, Gertz M, Nguyen GT, Scharfe M, Born I, Sippl W, Steegborn C, Schutkowski M. Chemical probing of the human sirtuin 5 active site reveals its substrate acyl specificity and peptide-based inhibitors. *Angew Chem Int Ed.* 2014;53(40):10728–10732.
208. Bheda P, Jing H, Wolberger C, Lin H. The substrate specificity of sirtuins. *Annu Rev Biochem.* 2016;85:405–429.
209. He B, Du J, Lin H. Thiosuccinyl peptides as Sirt5-specific inhibitors. *J Am Chem Soc.* 2012;134(4):1922–1925.
210. Ballatore C, Huryn DM, Smith 3rd AB. Carboxylic acid (bio)isosteres in drug design. *ChemMedChem.* 2013;8(3):385–395.
211. Singh MP, Wijeratne SSK, Zempleni J. Biotinylation of lysine 16 in histone H4 contributes toward nucleosome condensation. *Arch Biochem Biophys.* 2013;529(2): 105–111.
212. Leach FR, Yasunobu K, Reed LJ. Lipoic acid activation of the α-ketobutyrate oxidation system in cell-free extracts of *Streptococcus faecalis. Biochim Biophys Acta.* 1955;18 (2):297–298.
213. Tsunoda JN, Yasunobu KT. Mammalian lipoic acid activating enzyme. *Arch Biochem Biophys.* 1967;118(2):395–401.
214. Morikawa T, Yasuno R, Wada H. Do mammalian cells synthesize lipoic acid? *FEBS Lett.* 2001;498(1):16–21.
215. Khalil AM, Boyar FZ, Driscoll DJ. Dynamic histone modifications mark sex chromosome inactivation and reactivation during mammalian spermatogenesis. *Proc Natl Acad Sci USA.* 2004;101(47):16583–16587.
216. McNally MA, Hartman AL. Ketone bodies in epilepsy. *J Neurochem.* 2012;121(1): 28–35.
217. Allen BG, Bhatia SK, Anderson CM, Eichenberger-Gilmore JM, Sibenaller ZA, Mapuskar KA, Schoenfeld JD, Buatti JM, Spitz DR, Fath MA. Ketogenic diets as an adjuvant cancer therapy: history and potential mechanism. *Redox Biol.* 2014;2: 963–970.

218. Shibata T, Shimozu Y, Wakita C, Shibata N, Kobayashi M, Machida S, Kato R, Itabe H, Zhu X, Sayre LM. Lipid peroxidation modification of protein generates N^{ε}-(4-oxononanoyl) lysine as a pro-inflammatory ligand. *J Biol Chem.* 2011;286(22): 19943–19957.
219. Zhong H, Yin H. Role of lipid peroxidation derived 4-hydroxynonenal (4-HNE) in cancer: focusing on mitochondria. *Redox Biol.* 2015;4:193–199.
220. Lee SH, Rindgen D, Bible Jr RH, Hajdu E, Blair IA. Characterization of 2′-deoxyadenosine adducts derived from 4-oxo-2-nonenal, a novel product of lipid peroxidation. *Chem Res Toxicol.* 2000;13(7):565–574.
221. Lin D, Lee H-g, Liu Q, Perry G, Smith MA, Sayre LM. 4-Oxo-2-nonenal is both more neurotoxic and more protein reactive than 4-hydroxy-2-nonenal. *Chem Res Toxicol.* 2005;18(8):1219–1231.

CHAPTER THREE

Regulatory Effects of NAD^+ Metabolic Pathways on Sirtuin Activity

Ning Zhang, Anthony A. Sauve[1]
Weill Cornell Medical College, New York, NY, United States
[1] Corresponding author. E-mail address: aas2004@med.cornell.edu

Contents

Abstract

NAD^+ acts as a crucial regulator of cell physiology and as an integral participant in cellular metabolism. By virtue of a variety of signaling activities this central metabolite can exert profound effects on organism health status. Thus, while it serves as a well-known metabolic cofactor functioning as a redox-active substrate, it can also function as a substrate for signaling enzymes, such as sirtuins, poly (ADP-ribosyl) polymerases, mono (ADP-ribosyl) transferases, and CD38. Sirtuins function as NAD^+-dependent protein deacetylases (deacylases) and catalyze the reaction of NAD^+ with acyllysine groups to remove the acyl modification from substrate proteins. This deacetylation provides a regulatory function and integrates cellular NAD^+ metabolism into a large

Progress in Molecular Biology and Translational Science, Volume 154
ISSN 1877-1173
http://dx.doi.org/10.1016/bs.pmbts.2017.11.012

spectrum of cellular processes and outcomes, such as cell metabolism, cell survival, cell cycle, apoptosis, DNA repair, mitochondrial homeostasis and mitochondrial biogenesis, and even lifespan. Increased attention to how regulated and pharmacologic changes in NAD^+ concentrations can impact sirtuin activities has motivated openings of new areas of research, including investigations of how NAD^+ levels are regulated at the subcellular level, and searches for more potent NAD^+ precursors typified by nicotinamide riboside (NR) and nicotinamide mononucleotide (NMN). This review describes current results and thinking of how NAD^+ metabolic pathways regulate sirtuin activities and how regulated NAD^+ levels can impact cell physiology. In addition, NAD^+ precursors are discussed, with attention to how these might be harnessed to generate novel therapeutic options to treat the diseases of aging.

1. INTRODUCTION

Nicotinamide adenine dinucleotide (NAD^+) has been studied extensively for many decades, but continues to receive current research attention as new roles and functions of this cofactor have emerged in the regulation of protein modifications in cells. Through protein modification activity, while acting as a signaling substrate, it can regulate cellular metabolism and a diverse range of cellular events.[1,2] Specifically, NAD^+ is key to activities of NAD^+-consuming proteins, such as sirtuins,[3,4] poly(ADP-ribosyl) polymerases (PARPs),[5–7] and cyclic ADP-ribose synthases (CD38 and CD157),[8,9] These are NAD^+ consumers, which modify cellular protein and/or generate bioactive second messengers derived from NAD^+. These NAD^+ consumers link signaling directly to NAD^+ metabolism and it is now appreciated that profound biological consequences can be affected, such as life span and the development of age-associated disease syndromes.

NAD^+ metabolism provides a direct and potent means to regulate the activities of sirtuins.[10,11] The sirtuins are phylogenetically related to a founding member, called silent information regulator 2 (Sir2), which was first described in the budding yeast *Saccharomyces cerevisiae*. Characterized as an NAD^+-dependent deacetylase,[2] it takes its name from a role in heterochromatin formation and gene silencing.[12] Seven mammalian sirtuins (SIRT1–7) have homology to yeast Sir2 in the conserved catalytic core which binds acetyllysine and NAD^+ and is composed of 275 amino acids.[13] NAD^+ and acetyllysines react on the Sir2 active site to deacetylate lysine providing the most common catalytic and regulatory function of sirtuins. Later work revealed that some sirtuins could act as deacylases, where groups, such as succinyl and malonyl can be removed beyond the acetyl modification. NAD^+ dependence and a broad variety of targets enable sirtuins to regulate energy metabolism, stress response pathways, and associated transcriptional programs. Thus, NAD^+ directly links metabolic status with signaling and transcriptional activities occurring in cells.[14–16] These regulatory activities can fundamentally alter cellular processes and can be targeted by pharmacologic strategies designed to modulate NAD^+ metabolism.

In this chapter, an overview of the current state of knowledge of NAD^+ metabolism is presented and how NAD^+ metabolic pathways can regulate sirtuin activities. Topics include NAD^+ metabolism, biological consequences of NAD^+ signaling through sirtuins, cellular compartmentalization of sirtuins and compartmentalized NAD^+ metabolism, and how this compartmentalization can impact sirtuins. In addition, an emphasis will be given to recent studies that attempt to show how NAD^+ precursors can raise intracellular NAD^+ levels and provide accentuated sirtuin action. A consideration of how NAD^+ precursors might represent a new class of potential therapeutics for treatment of metabolic disorders and age-related diseases is also provided.

2. NAD^+ STRUCTURE AND CHEMISTRIES

NAD^+ incorporates a low pK_a electron deficient pyridine (conjugate acid form), in the form of nicotinamide, and the pyridinium is redox active, as well as prone to substitution at the ribose carbon of the pyridine C–N bond. This substitution propensity is due to the excellent leaving group properties of the pyridine base. These chemical properties account for the biological effects and chemical activities of NAD^+. Due to its abundance and centrality to cellular metabolism, NAD^+ is well suited to regulate catabolic and biosynthetic pathways and to influence a number of cellular events, including transcription, signaling, and cell survival.[17,18]

2.1 NAD^+ as an Enzyme Cofactor

NAD^+ in its most commonly known form is a hydride-transfer acceptor, serving a wide variety of metabolic transformations, as it interconverts between its oxidized form (NAD^+) and reduced form (NADH). Reduction is achieved via addition of saturation at position 4 of the pyridine

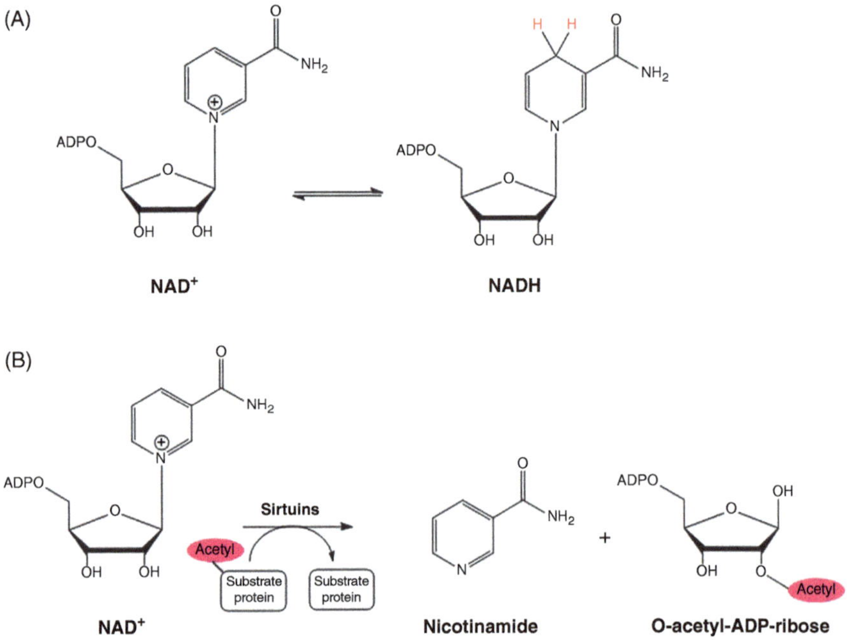

Fig. 1 Nicotinamide adenine dinucleotide (NAD^+) can be used as both coenzyme and cosubstrate. (A) Transfer of hydride to nicotinamide of NAD^+ to form its reduced form NADH. (B) NAD^+-dependent reaction catalyzed by sirtuin enzymes. Upon deacetylation reactions, sirtuins catalyzes the transfer of acetyl group from lysine residues on protein substrates to the ADP ribose moiety of NAD^+, followed by releasing nicotinamide, O-acetyl-ADP ribose, and deacetylated substrates.

ring (Fig. 1A).[19] NAD^+ directly participates in compartment-specific but central metabolic pathways, such as glycolysis, pyruvate dehydrogenase, the tricarboxylic acid (TCA) cycle, fatty acid oxidation, amino acid oxidation, and oxidative phosphorylation. In glycolysis, NAD^+ is converted to NADH in the glyceraldehyde-3-phosphate dehydrogenase step.[20] This process occurs in reverse for gluconeogenesis in liver, and so NAD^+ involvement is crucial for catabolism, as well as biosynthetic activities relating to glucose interconversion with pyruvate. In the mitochondrial compartment, NAD^+ is converted to NADH at three distinct steps in the TCA cycle in which acetyl-CoA is oxidized to produce the bulk of mammalian carbon dioxide. The NADH form is subsequently oxidized through a series of redox reactions in oxidative phosphorylation to provide the bulk of cellular energy to produce ATP.[21,22] The ratio of NAD^+ to NADH, which is typically 10:1

reflects the strong catabolic role of NAD^+ as an oxidant. Cellular homeostasis maintains this ratio in the strongly oxidizing redox state. This ratio can be considered as a readout of instantaneous metabolic conditions within cells and tissues.[23,24]

NAD^+ is also a precursor substrate for $NADP^+$ and NADPH, which participate in biosynthesis and reactive oxygen detoxification. As compared with NAD^+ (NADH), these dinucleotides are present at less abundant levels, perhaps at 10% of cellular NAD^+ content.[25] As opposed to NAD^+/(NADH) the ratio $NADP^+$/NADPH is maintained such that NADPH is typically well in excess to $NADP^+$, with the ratio much less than 1. Thus, the cell maintains two opposed redox pairs with the $NADP^+$/NADPH pair in a reducing state. This reductive poise is consistent with the role of NADPH in biosynthesis and oxygen detoxification where reducing equivalents are required. NADPH is a key reducing substrate to convert oxidized glutathione to reduced glutathione. Reduced glutathione is a key protectant in cells to resist the toxicity of reactive oxygen species. A key feature of the redox chemistry of NAD^+ and NADH is that it does not change net NAD^+/NADH level in cells, thus, this process requires no net synthesis of NAD^+ to maintain homeostasis.

2.2 NAD^+ as an Enzyme Cosubstrate for ADP-Ribosyl Transferase and Sirtuins

On the other hand, NAD^+ is actively consumed during the enzymatic processes of ADP-ribosyl transferase exemplified by the reaction of sirtuins. Sirtuins are ADP-ribosyl transfer-type enzymes, which catalyze a protein lysine deacylation reaction releasing nicotinamide (NAM), an enigmatic product O-acyl ADP-ribose, and the deacylated substrate (Fig. 1B).[26,27] NAD^+ serves as the donor of ADP-ribose in the reaction, resulting in consumption. ADP-ribose conjugated to nicotinamide in NAD^+ activates ADP-ribose to transfer to the acyl oxygen, facilitated by the sirtuin active site, which binds acylated lysines and places them in proximity to the α-face of this labile bond. The displacement of nicotinamide by the acyl-oxygen occurs via an ADP-ribosyl oxacarbenium-like transition state.[28] This reaction forms an enzyme-bound ADP-ribosyl-peptidylimidate,[26] which subsequently hydrolyzes to form the observed products. ADP-ribosyl transfer by NAD^+ is a general reaction used to modify proteins (deacetylation by sirtuins or ADP-ribosylation) and is of fundamental importance for shaping cellular physiology.[29]

3. INTEGRATION OF NAD^+ METABOLISM INTO SIGNALING PATHWAYS VIA SIRTUIN ACTIVITY

Recognition that NAD^+ can serve roles beyond its typical redox functions has continuously expanded since the 1960s with discovery of poly (ADP-ribosyl) polymerases, mono (ADP-ribosyl) transferases and finally the discovery of sirtuins. Currently, it is understood that NAD^+ is a vital and central signaling metabolite, in addition to being a vital and central redox substrate. NAD^+ in signaling appears to be a key regulator of acylation status of protein lysines on sites subject to modification, and notably, NAD^+ signaling is a strong component in regulation of a number of key and central transcriptional programs that adapt and shape cell physiology.[30,31] In addition to the understood role of NAD^+ in sirtuin activity, the role of AADPR is still not well understood. However, NAD^+ and NAD^+ phosphate metabolites, such as cyclic ADP-ribose and nicotinic acid adenine dinucleotide phosphate (also called NAADP), have emerged as key second messengers in Ca^{2+} signaling.[32,33] These elucidated roles of downstream metabolites of NAD^+ suggest still undiscovered second messenger activities of AADPR in cell signaling.[34]

A renewed interest in NAD^+ metabolism, directed at understanding the pathways it serves and regulates, has been extended to understanding NAD^+ metabolism itself. For example, it has become increasingly relevant to ask if concentrations of NAD^+ and other related metabolites are tuned to match cell needs. The answer to this question has been an unequivocal affirmative, that NAD^+ is a highly regulated yet dynamic metabolite.

NAD^+ levels are in fact modulated by dynamic changes in NAD^+ biosynthesis and consumption. Consumption is increased during increased cellular stress, which can increase rates of NAD^+ turnover. Specifically increased consumption can occur with activation of PARP enzymes, which can become hyper activated by genotoxicity. Nevertheless, base-line NAD^+ turnover is robust, and the half-life of intracellular NAD^+ is estimated to be ~1–4 h,[35] and its typical concentrations are estimated to be between ~0.2 and 0.5 mM.[26] On the other hand, dynamics in NAD^+ biosynthesis can strongly influence cellular NAD^+ levels as shown by our lab and others. For example, Sartorelli and coworkers[36] identified AMPK activation in response to glucose restriction as an impetus for increased transcription and stabilization of the NAD^+ biosynthetic enzyme nicotinamide phosphoribosyltransferase (Nampt), causing increases in cellular NAD^+ levels in C2C12 myotubes. Yang *et al.*[37] showed that fasted animals exhibit a similar response leading to increased Nampt transcription and protein levels, resulting in robust increases in cellular NAD$^+$. Nampt appears to be limiting for NAD$^+$ biosynthesis in many tissues and is a key regulator of NAD$^+$ concentration. This fact is partly explainable by a low μM K_m for nicotinamide on this enzyme,[38] so that nicotinamide is typically not limiting. Exercise also regulates Nampt transcription, possibly through induction of AMPK, and the general idea that exercise and low calorie diets initiate programs to upregulate NAD$^+$ biosynthesis in mammals[39,40] is now fairly well established.[41] Demonstrating that regulated changes in NAD$^+$ levels lead directly to changes in activities of sirtuins and other downstream NAD$^+$ consumers has been a key goal.

Thus, a well-known hypothesis is a proposal of the sirtuin family of deacylases as key players in a metabolic monitoring system integrated into sensing NAD$^+$ dynamics. This hypothesis leads directly from the substrate dependence of these enzymes, wherein the bioavailability of NAD$^+$ becomes a highly relevant parameter to their activity and function. The coupling of protein deacylation activity responsive to NAD$^+$ concentration, in principle can link to cellular NAD$^+$/NADH redox state, and overall NAD$^+$ bioavailability to protein regulation.[31] Sirtuins are central mediators of cellular adaptation to nutritional stress. Thus, NAD$^+$ bioavailability could in principle generate metabolic adaptations, where sirtuins drive programs leading to, for instance, enhanced mitochondrial biogenesis, suppressed lipogenesis, and modulation of respiration.

A fundamental and reasonable basis for conceptual linkage of sirtuin signaling intensities to NAD$^+$ levels is a consideration of the biochemical parameters of the enzymes themselves. Published data indicate that sirtuins have K_m values for NAD$^+$ in the range 50–500 μM, which overlaps published NAD$^+$ concentrations within mammalian cells (0.3–0.6 mM, see earlier). The K_m of SIRT1 is estimated to be in the range of 50–500 μM,[26,42] and the K_m of SIRT2 is measured to be ~100 μM.[43] For the mitochondrial sirtuins, the K_m of SIRT3 for NAD$^+$ is ~300 μM,[44] and the K_m of SIRT5 is ~100 μM.[45] Moreover, the effective K_m is modified by the presence of NAM, as NAM acts as a mixed-inhibitor, which increases the "apparent K_m." In addition to biochemical data, there is impressive data supporting changes in NAD$^+$ levels in each subcellular compartment leading to direct regulation of sirtuin activity.[46] Both nucleocytoplasmic and nuclear changes in NAD$^+$ concentrations accompany responses to physiologic stresses, such as fasting and low calorie diets.[47,48] This sets up scenarios wherein stresses can modulate sirtuin activities at the subcellular level by linkage through NAD^+ dynamics. This also occurs negatively, wherein DNA damage elicits declines in NAD^+ level, to reduce sirtuin activity as reviewed recently by our lab.[41] Moreover, in both genetic and dietary models of obesity, metabolic dysfunction is correlated with a reduction in NAD^+ content in multiple tissues.[49,50] Evidence for reduced sirtuin activity in physiologic situations that reduce NAD^+ levels affirm the connection between NAD^+ and sirtuin activity.[51]

4. MAMMALIAN SIRTUINS ARE COMPARTMENT-SPECIFIC

Mammalian cells maintain subcellular compartments, which are involved in distinct cellular functions, such as energy creation, and genetic maintenance and activity, and subcellular distributions of sirtuin isoforms to these compartments highlights the specialization of function inherent to the distinct mammalian sirtuin isoforms. NAD^+ activity is modulated in these separate compartments, with separate NAD^+ biosynthetic mechanisms featured in the different compartments. The seven members of the sirtuin family in mammals SIRT1–7[52] distribute to the different cellular compartments with SIRT1, SIRT6, and SIRT7 predominantly nuclear;[53] whereas SIRT3, SIRT4, and SIRT5 are mitochondrial.[54] SIRT2, a somewhat more distributed sirtuin, resides in the cytoplasm. Although all sirtuins share a NAD^+-binding catalytic domain and catalyze chemically similar reactions, specific sirtuins residing and acting within subcellular compartments regulate diverse and separate processes, including glucose homeostasis, gluconeogenesis, TCA cycle, mitochondrial biogenesis, insulin secretion, adipogenesis and adipolysis, apoptosis, senescence, and metabolism (Table 1).

The compartmentalization of the sirtuins is not completely static and displays dynamic characteristics, with shuttling between cellular compartments. For example, SIRT1 can shuttle between nuclear and cytoplasmic compartments and the nucleocytoplasmic shuttling is regulated by nuclear localization signals and nuclear export signals in the amino acid sequences of SIRT1.[98] SIRT2 can also regulate gene expression by deacetylating transcriptional regulators that display nucleocytoplasmic shuttling.[99] SIRT3 has also been reported to translocate between the mitochondria and nucleus in response to cellular stress.[53]

Table 1 Summary of the Localization, Enzymatic Activity, and Biological Function of the Seven Mammalian Sirtuin Isotypes (SIRT1–7).

Isotypes	Locations	Enzymatic Activities	Biological Functions
SIRT1	Nucleus Cytoplasm	Deacetylation[55]	DNA repair,[56,57] tumor suppression,[58] apoptosis,[59] metabolism,[60] cell cycle,[61] longevity,[62] insulin signaling[63]
SIRT2	Cytoplasm Nucleus	Deacetylation[64] Demyristoylation[65]	Cell cycle progression,[66] genome stability,[66] adipogenesis,[64] longevity,[67] autophagy,[68] immune response,[69] bacterial infection[70]
SIRT3	Mitochondria Nucleus	Deacetylation[71] Decrotonylation[72]	Stress resistance,[73] TCA cycle,[74] fatty acid oxidation,[75] tumor suppression,[76] cell survival[77]
SIRT4	Mitochondria	ADP-ribosylation[78] Deacetylation[79]	Cell cycle,[80] tumor suppression,[81,82] fatty acid oxidation,[83] insulin secretion[78,84]
SIRT5	Mitochondria	Demalonylation[85] Desuccinylation[85] Deacetylation[45]	Fatty acid oxidation,[86] ketogenesis,[87] stress resistance,[88] urea cycle[45]
SIRT6	Nucleus	Deacetylation[89] ADP-ribosylation[90]	DNA repair,[91] protein secretion,[92] longevity[93]
SIRT7	Nucleus	Deacetylation[94]	Genome stability,[95] transcriptional regulation,[96] tumor suppression[97]

TCA, Tricarboxylic acid.

5. COMPARTMENTALIZED NAD$^+$ DYNAMICS AND EFFECTS ON SIRTUIN ENZYMES IN MAMMALIAN CELLS

Compartmentalization of NAD$^+$ has been difficult to elucidate but essential to understand, given that sirtuins act locally within distinct subcellular loci modulated by NAD$^+$ bioavailability.[100]

A key fact is that NAD$^+$ is not evenly distributed in subcellular compartments, that is, nucleus, cytosol, and mitochondria. Several studies report that mitochondrial NAD^+ concentrations are higher than in other compartments.[101] It was shown that in mouse skeletal muscles, the mitochondrial NAD^+ level was found to be approximately twofold higher than the rest of the cell.[102] Additional studies indicate that mitochondrial NAD^+ concentration are ~250 μM in cultured HEK293 cells,[37] whereas nuclear NAD^+ concentration might be ~70 μM,[45,103] and the nuclear NAD^+ levels might be lower than cytosolic NAD^+ levels.[104] Thus, it is apparent that different cellular compartments could have different NAD^+ pools with different subcellular concentrations. Evidence also suggests that depletion of these pools is differentially sensitive to cell stresses. For example, DNA strand breaks temporarily expose the nuclear NAD^+ pool to rapid turnover, with less apparent in cytoplasm, and mitochondria have very limited PARP activity, so NAD^+ turnover under genotoxic stress is less rapid there.[37]

The key major components in the biosynthesis of NAD^+ are the NAM salvage pathway (Fig. 2), which synthesizes NMN from NAM using Nampt.

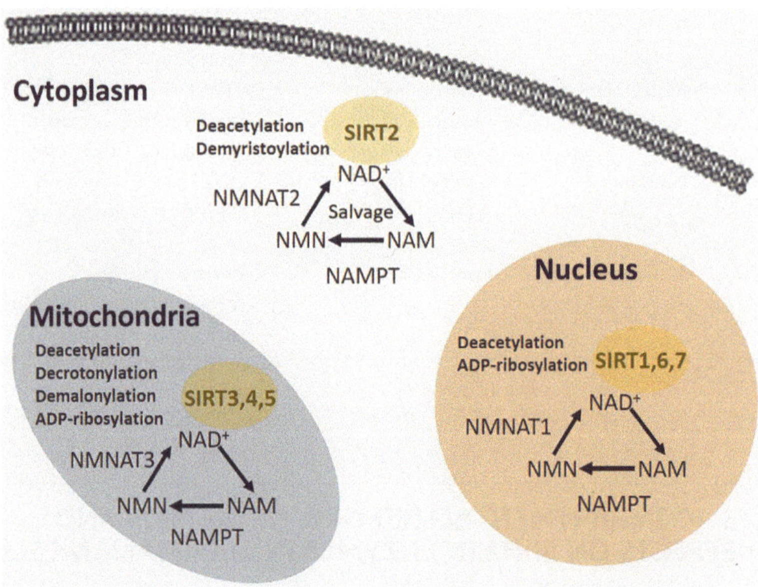

Fig. 2 NAD^+ metabolism in cellular compartments and localization, enzymatic activity, and modulation of sirtuins. SIRT1, SIRT6, and SIRT7 are localized primarily to the nucleus; whereas SIRT3, SIRT4, and SIRT5 are present in the mitochondria. SIRT2 is largely localized in the cytosol. NAD^+ salvage pathway recycles NAD^+ from NAM by virtue of nicotinamide phosphoribosyltransferase (NAMPT1–3). *NAM*, Nicotinamide; *NMN*, nicotinamide mononucleotide.

The downstream enzymes are the adenylyltransferases NMNAT1–3. Three NMNAT isoforms have been identified (NMNAT1–3), which vary in their catalytic properties and subcellular localization.[105] NMNAT1 is nuclear, NMNAT2 is enriched in cytoplasm, and NMNAT3 is mitochondrial. It has been determined that the enzyme Nampt is subject to dynamic regulation and is the most responsible enzyme for NAD^+ homeostasis in mammalian cells in mammals.

Some new technical advances to elucidate the levels of subcellular NAD^+ and how they are affected by perturbations have arisen in the form of genetically encoded NAD^+ sensors. These genetically encoded sensors can be localized to distinct cellular compartments to provide measurement of NAD^+ concentrations in real time. Perhaps the best example of this type of sensor is those developed by Cambronne *et al.*[106] The prototype sensor is comprised of a circularly permuted fluorescent Venus protein conjugated to a bacterial ligase, which uses NAD^+ as a substrate. This probe was developed with the ability to target it to mitochondrial, cytoplasmic, or nuclear loci. These workers determined that they could assay the absolute concentrations of NAD^+ in HEK293 cells, and were able to determine that mitochondrial free NAD^+ is maintained at approximately twofold higher (200 μM) than cytoplasmic or nuclear compartments, that nuclear and cytoplasmic levels (100 μM) are about the same. Interestingly, with application of Nampt inhibitor FK866, it was observed that cytoplasmic and NAD^+ pools were reduced much quicker than mitochondrial NAD^+ pools, ($t_{1/2} = 2$ h vs. 4–6 h for mitochondria) suggesting different rates of turnover or different mechanisms of homeostasis. In addition, these workers found that distinct NAD^+ biosynthetic mechanisms operate to control local NAD^+ levels, with NMNAT1 providing support for NAD^+ in nucleus; whereas NMNAT2 provides support for both nuclear and cytoplasm. Both NMNAT2 and NMNAT3 appear to influence mitochondrial NAD^+ pools. The fact that NMNAT2, which is cytoplasmic can alter mitochondrial NAD^+ pools suggests that there may be an NAD^+ transporter to conduct NAD^+ into the mitochondrial compartment. This putative transporter is currently unidentified, but NAD^+ transporters have been identified in bacteria, yeast, and plants. These initial studies appear to confirm that free NAD^+ concentrations in distinct subcellular compartments are differentially regulated, and that a significant fraction of total NAD^+ exists in a protein bound form, thereby reducing the available NAD^+ pool. These data also reinforce the concept that under most circumstances, local NAD^+ concentrations are subsaturating for sirtuin activity, and dynamics in NAD^+ level would be expected to alter subcellular sirtuin activity. How various cell types and different cellular perturbations alter these subcellular NAD^+ distributions is just beginning to be elucidated.

6. MAJOR BIOLOGICAL FUNCTIONS OF SIRT1 AND SIRT3 VIA COMPARTMENT-SPECIFIC NAD CHANGES

6.1 Nuclear Effects

SIRT1 regulates a broad panel of substrates by deacetylation and this includes tumor suppressors, such as p53, DNA repair genes,[56,107] apoptotic products, such as Bax,[108] PPARγ which regulates adipogenesis,[109] PGC1α which regulates mitochondrial biogenesis,[110] androgen receptor,[111] and even NF-κB[110] which regulates inflammation. Thus, among sirtuins it is one of the most studied and has the most pleiotropic effects. Evidence that NAD^+ regulates SIRT1 activity provides one of the most compelling cases for the argument that sirtuin activity is linked to dynamics in NAD^+ metabolism. This can be manifested in different tissues, giving rise to different biological effects, and even the specific targets can change dependent on context.

One of the earliest examples of this type of regulatory mechanism was identified by Puigserver and coworkers, investigating roles of SIRT1, NAD^+, and PGC1α in liver.[112] In liver, PGC1α serves to regulate transcription of genes that promote gluconeogenesis.[113] This function of PGC1α is stimulated by fasting or low glucose conditions, and by cAMP signaling to the liver. Puigserver and coworkers noted that PGC1α activity is substantially modulated by SIRT1 activity, which deacetylates PGC1α, leading to a hypoacetylated and activated form. Coincident with the fasting and low glucose state, these authors noted NAD^+ levels increased, leading to increased activation of PGC1α and gluconeogenesis.[60]

Interestingly, recent work on this system has looked at pathophysiologic depletion of NAD^+ and corresponding destabilization of PGC1α and reduced gluconeogenesis caused by dioxin activation of aryl hydrocarbon receptor. Rifkind and coworkers[114] showed that increasing NAD^+ levels in liver, by treatment with either nicotinamide or nicotinamide riboside could cause increased stabilization of PGC1α, SIRT1 activity, and importantly restored gluconeogenesis. Thus, in liver, SIRT1 activity appears to be a potent modulator of PGC1α and gluconeogenesis that is coupled to NAD^+ levels, and this has roles in normal physiology, as well as in abnormal physiology.

Another interesting regulatory system involving dynamic regulation of NAD^+ levels and coordinated SIRT1 activity has been identified in circadian rhythms. Mammalian systems are subject to the circadian clock, which acts at the level of transcription and is controlled by a set of clock gene products, such as BMAL, PER, etc. A large subset of mammalian genes has been shown to have circadian transcription patterns that interface with the CLOCK system. Independently, Sassone-Corsi and coworkers found that SIRT1 and NAD^+ levels regulate the clock system via an interesting mechanism whereby NAD^+ levels oscillate, via transcriptional and translational oscillation of Nampt activity, to control SIRT1 activity.[115–117] SIRT1 activity deacetylates BMAL1 opposing the acetyltransferase activity of CLOCK. Thus, oscillation of NAD^+ levels drives oscillatory SIRT1 activity, which can drive circadian behavior by circadian modulation of BMAL1.

Another direct regulator of SIRT1 via NAD^+ synthesis is the adenylyl transferase NMNAT1. Kraus and coworkers examined the role of coordinated synthesis of NAD^+ effected by NMNAT1 activity with activation of SIRT1 activity at chromatin.[104,118] NMNAT1 has been found to reside predominantly in the nucleus, where it makes a stable interaction with SIRT1. In the initial studies Zhang *et al.* found that hairpin knockdown of either Nampt or Nmnat1 led to declines in NAD^+ level, leading to approximately 200 total gene transcripts that are significantly upregulated or downregulated in each case (log2 difference > 0.5 or log2 difference < -0.5) with about 20% of them in common. In the cases where gene transcripts were found in common, knockdown of SIRT1 provided a similar pattern, based on correlation of transcript change to either NMNAT1 or Nampt, indicating that all three of these factors regulate a subset of total genes similarly, suggesting they participate together in this process. A specific feature of the regulation is that SIRT1 colocalizes with NMNAT1 to target promoters, and it is suggested that NMNAT synthesis of NAD^+ in the vicinity of the catalytic site may assist SIRT1 activity. Interestingly, NMNAT1 also forms an association with PARP1 at promoters and may play a similar role.

In models of skeletal muscle the effects of NAD^+ on SIRT1 activity were first well studied by Sartorelli and coworkers, who showed that C2C12 myotubes increase Nampt activity upon restriction of glucose.[36] Interestingly, these authors showed that the increased NAD^+ levels and the concomitant increased SIRT1 activity leading to decreased differentiation of myoblasts from mice. Importantly, this is a key paper for placing AMP kinase activation upstream of upregulation of Nampt. AMP kinase is a key nutrient and energy sensor, which is a key player in skeletal muscle physiology. Key experiments in SIRT1$^{+/-}$ heterozygote myoblasts showed that SIRT1 deficient myoblasts did not respond to glucose restriction or AMP kinase activation using AICAR as a pharmacological tool.

Interestingly, AMP kinase is a key player in fuel switching in skeletal muscle and drives PGC1α activation. PGC1α is a key player driving mitochondrial biogenesis. As PGC1α activation requires SIRT1 activity as described earlier, the possibility that increased NAD$^+$ may be sufficient to achieve activation of PGC1α in skeletal muscle was examined by Auwerx and coworkers.[47] They found that global knockout of PARP1 caused impressive increases in NAD$^+$ levels in skeletal muscle in mice. These increases in muscle NAD$^+$ led to demonstrable increases in mitochondria as measured by succinyl dehydrogenase staining, and western blots of mitochondrial complex proteins. Evidence for activation of SIRT1 was obtained in the form of increased deacetylation of FOXO1 and PGC1α in knockouts. This genetic proof of concept for increased NAD$^+$ leading directly to mitochondrial biogenesis has been further explored in a number of intriguing therapy-directed scenarios, as covered in a later section in this chapter focused on NAD$^+$ precursors and pharmacology of NAD$^+$ manipulation.

Finally, recent work out of the Sinclair laboratory places SIRT1 as an essential player in regulation of mitochondrial fitness, linked to normal but age-vulnerable NAD$^+$ homeostasis.[119] A key aspect of this recent work is the demonstration that declines in tissue NAD$^+$ level occur with advancing age, and that this gradual depletion of cellular NAD$^+$ results in loss of mitochondrial fitness, due to imbalances in nuclear-encoded and mitochondrial-encoded proteins. In other words, impaired mitochondrial fitness can be induced through the development of incorrect stoichiometry of nuclear-encoded and mitochondrial proteins which are required for the correct assembly of mitochondrial structures. Thus, declines in NAD$^+$ level mediated by commensurate declines in sirtuin signaling can result in functional deficits, such as decreased respiratory capacity and reduced ATP production. Repletion of NAD$^+$, via NMN, can restore mitochondrial structure and protein ratios, leading to increased ATP production. Thus, a key type of age-dependent decline in mammalian tissues may be due to weakness in NAD$^+$-mediated signaling through sirtuins, causing serious consequences in energy physiology and organism fitness.

6.2 Mitochondrial Effects

Evidence indicates that SIRT3 is the dominant mitochondrial deacetylase.[120] There are hundreds of acetylation sites that have been identified for mitochondrial proteins and some of these have been determined to have regulatory significance. Mitochondrial acetylation levels of proteins are quite dynamic, increasing substantially with SIRT3 deletion[121]; however, more relevantly, responding to cellular stimuli, such as nutrient intake. This intake can affect NAD^+ synthesis, and changes in NAD^+ levels in mitochondria is an important mechanism for control of sirtuin activity in the mitochondria.

Mitochondrial NAD^+ levels fluctuate under a variety of circumstances. As already mentioned, Cambronne *et al.*[106] showed that NMNAT2 and NMNAT3 levels can influence mitochondrial NAD^+ levels. Sinclair and coworkers[37] showed that Nampt expression levels directly impact mitochondrial NAD^+ levels. In particular, in this work, mitochondrial NAD^+ levels were found to be a limiting factor for SIRT3 and SIRT4 activity, wherein reduced NAD^+ levels caused genotoxin-treated cells to undergo apoptosis; whereas Nampt overexpression proved to be protective, in a background of total depletion of cytoplasmic NAD^+.

In a study with an intriguing connection to biological aging, NAD^+ depletion and mitochondrial SIRT3 was uncovered by Denu and coworkers,[122] who determined that C57/B6 mice become deaf with age, but that calorie restricted mice are protected. $SIRT3^{-/-}$ mice were not able to experience the benefit of calorie restriction, indicating the necessity of SIRT3 in this effect and implicating declines in mitochondrial NAD^+ as a possible origin of the wild type phenotype on normal diet. This impressive study placed mitochondrial NAD^+ as a central weakness in aging that could lead to organ dysfunction and loss of quality of life. The declines in tissue NAD^+ in aged mice has since been robustly confirmed by Auwerx, Sinclair, and Chini groups.[119,123,124]

So, if NAD^+ is a limiting regulator of SIRT3, what are the biological effects of SIRT3 to give rise to such profound effects? Among substrates of SIRT3 are acetyl-CoA synthetase, superoxide dismutase 2,[125] isocitrate dehydrogenase 2,[126] and cyclophilin D,[127] which regulates the mitochondrial permeability transition. SIRT3 is responsive to stress-regulating mitochondrial antioxidant manganese superoxide dismutase in microglia.[125] It appears that SIRT3 effects not only increase activity of a key detoxification enzyme, but also increase activity of enzymes that produce reduced NADPH, which is a key substrate to improve reduced glutathione/oxidized glutathione ratio. SIRT3 inhibits mitochondrial reactive oxygen species (ROS) production but also deacetylates mitochondrial proteins, regulating proliferation, differentiation, and survival.[71,128,129]

The activity of SIRT3 as a potential survival factor, which is sensitive to NAD^+ concentrations in mitochondria has been investigated in an interesting study using the NAD^+ precursor nicotinamide riboside (NR). Jaffrey and coworkers[130] showed that overexpression of SIRT3 could confer protection to noise induced hearing loss, in a study reminiscent of the study of Denu and coworkers.[122] In this instance, the hearing loss was induced by trauma and not age. Administration of NR, which increases mitochondrial NAD^+ content led to nearly complete protection of auditory neurons and functional preservation of hearing. This study highlighted the profound activity of SIRT3 as a possible neuroprotectant factor and pointed to increased mitochondrial NAD^+ through NAD^+ enhancers as a means to activate SIRT3. The data suggests SIRT3 is typically limited by protein abundance and NAD^+ level in mitochondria. Data from Canto *et al.*[50] confirms that NR administration enhances mitochondrial NAD^+ levels well above control levels, and that NR causes increased deacetylation of SOD2. The Jaffrey study indicates that NR can mediate protective effects by enhanced NAD^+ availability in the mitochondria of neurons via activation of SIRT3.

As aging occurs, declines of NAD^+ are expected to weaken SIRT3 activity, as already mentioned. A number of studies suggest weakened SIRT3 activity are involved in tumor progression. SIRT3 expression is reduced in human breast cancers, and its loss correlates with the upregulation of HIF1α target genes,[131] and SIRT3 deficiency results in a defect in cellular iron homeostasis and many cancer cells exhibit dysregulation in iron metabolism.[132] These factors suggest that it is worthwhile to explore if weakening NAD^+ homeostasis has carcinogenic consequences, in light of these findings. Weakened SIRT3 activity with age may also have important consequences in development of neurodegenerative syndromes.

6.3 Regulation of Yeast Sirtuins by NAD^+ Metabolism

The founding member of the sirtuin family, Sir2, was first identified in yeast[133] and was the founding member of the NAD^+ deacetylase family of enzymes. The yeast Sir2 has a major role in transcriptional silencing, within the TEL, rDNA, and MAT loci, which are subject to heterochromatic silencing. Sir2 in this role also acts as a critical determinant of replicative lifespan in this organism.[134–136] Sir2 longevity effects in yeast were extrapolated experimentally to metazoans, such as *Caenorhabditis elegans* and *Drosophila melanogaster*,[42,137] which live longer with increased Sir2 gene dosage. Additional Sir2 homologs in yeast include, Hst1, Hst2, Hst3, and Hst4.[138] Hst1 interacts with another transcription regulator, the budding yeast protein Sum1, in a gene-specific manner to regulate transcription.[139] Hst2 is the most abundantly expressed sirtuin in yeast and exhibits a major fraction of measurable intracellular NAD^+-dependent deacetylase activity.[140] Hst2 coordinates with Hst1 to downregulate subtelomeric gene transcription.[141] Hst2 has been shown to exhibit some redundancy with Sir2 and can contribute to rDNA silencing[142] and repress recombination at this locus.[136] Unlike Sir2, Hst2 localizes primarily in the cytoplasm and shuttles into the nucleus likely accounting for a weaker phenotype upon deletion.[143] In fact, individual deletion of these homologs does not interfere with gene silencing, although strong effects are observed with deletion of Sir2. Hst3 and Hst4 deleted together decrease telomeric silencing and cell cycle progression, with accentuated genome instability.[138] This instability and loss of silencing effect is linked to increased acetylation of H3K56, which is involved in silencing and formation of heterochromatin. In yeast, the Sir2 family is intimately involved with chromatin modification,[144] much more than in mammalian cells. Explicitly, Sir2, Hst1, Hst2, Hst3, and Hst4 appear to function as histone deacetylases specific for a variety of histone modifications that affect cell cycle progression, transcriptional silencing, and genome stability. These activities are subject to regulation by NAD^+ metabolism. NAD^+ and nicotinamide are the principle metabolites regulating sirtuins in yeast, the former acting as a key activator, and the latter acting as an inhibitor.

Yeast sirtuins are metabolic sensors, which respond to physiological changes in NAD^+ and NAM levels. There are two major NAD^+ biosynthetic pathways in yeast (Fig. 3). One involves de novo synthesis using tryptophan via 6 enzymatic steps (Bna1-Bna6), which convert tryptophan to quinolinic acid (QA). QA is converted to nicotinic acid mononucleotide (NaMN) by quinolinic acid phosphoribosyltransferase (QAPRT).[145] NaMN is a shared metabolite of both NAM and nicotinic acid (NA) recycling pathway and the de novo pathway.

NA is metabolized by the yeast version of the Preiss–Handler pathway[146,147] that starts from NA, which can be internalized into cells by the NA permease Tna1. The first reaction produces NaMN by coupling of NA and PRPP catalyzed by nicotinic acid phosphoribosyltransferase (Npt1). In yeast, there is no Nampt equivalent, and NAM is also metabolized to NaMN.

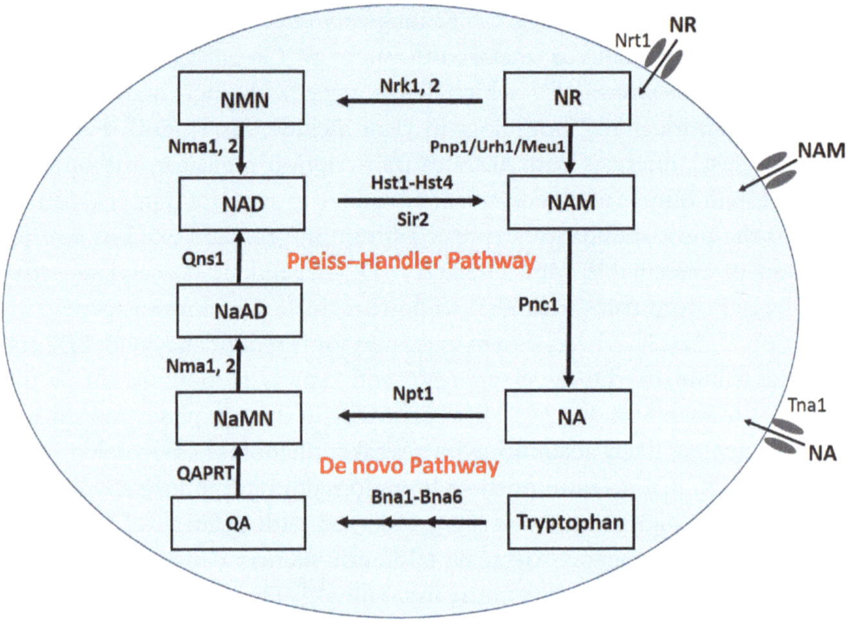

Fig. 3 The model of NAD^+ biosynthesis and metabolism in yeast. NAD^+ is synthesized via de novo pathway using tryptophan as the starting substrate or the Preiss–Handler pathway using NAD^+ precursors NA, NAM, and NR. The sirtuins (Hst1-Hst4 and Sir2) in yeast produce NAM during the deacetylation reaction. *NA*, Nicotinic acid; *NaAD*, nicotinic acid adenine dinucleotide; *NAM*, nicotinamide; *NaMN*, nicotinic acid mononucleotide; *NMN*, nicotinamide mononucleotide; *Npt1*, nicotinic acid phosphoribosyltransferase; *NR*, nicotinamide riboside; *Pnc1*, pyrazinamide/nicotinamidase carboxyhydrolase 1; *QA*, quinolinic acid, *QAPRT*, quinolinic acid phosphoribosyltransferase.

The first step in this process is hydrolysis of nicotinamide to NA by the enzyme pyrazinamide/nicotinamidase carboxyhydrolase 1 (Pnc1). Thus at least three different metabolic precursors are converted to NaMN in yeast, namely tryptophan, NA and nicotinamide.

NaMN proceeds to NAD^+ by adenylation by redundant adenylyltransferases Nma1 or Nma2 to form nicotinic acid adenine dinucleotide (NaAD). The acid dinucleotide is subsequently amidated via the NAD^+ synthetase Qns1 to form NAD^+ in an ATP dependent reaction.

Studies by labs of Smith and coworkers[148] and Sinclair and coworkers[149] examining Sir2 activities in nicotinamide supplemented media show that modulated activity of the nicotinamidase (encoded by *PNC1*) can have dramatic consequences on Sir2 mediated silencing and lifespan. Interestingly, *pnc*Δ strains do not have defects in NAD^+ level, suggesting that Npt is typically operating at saturation in these strains (due to sufficient NA in media), but *pncΔ* strains are defective in Sir2-mediated silencing and have shortened replicative lifespan. Overexpression of *PNC1* gives the opposite phenotype, namely, they do not show increased NAD^+ levels but they have significantly extended lifespan and improved gene silencing. These genetic and metabolic results implicate nicotinamide as a powerful negative regulator of Sir2 in yeast. Nicotinamide is a robust inhibitor of yeast Sir2 enzymatic activity.[150,151]

Effects of NAD^+ can be realized by manipulation of an independent salvage pathway involving NR. The nucleoside NR can be imported by the NR transporter Nrt1 and phosphorylated by one of two nicotinamide riboside kinases (Nrk1, 2) to form nicotinamide mononucleotide (NMN),[152] which can be adenylated by Nma1/Nma2 to produce NAD^+. NR can also be degraded into NAM via hydrolysis by three nucleoside-splitting enzymes (Pnp1/Urh1/Meu1)[153] leading to the synthesis of NaMN as already described.

The most crucial enzymes in the yeast NAD^+ synthesis, for maintenance of NAD^+ homeostasis, appear to be those involved in the recycling of NA, as deleting the enzyme Npt1 causes a significant decrease in total cellular NAD^+ level, and negatively impacts Sir2-mediated silencing in *S. cerevisiae.*[154,155] Interestingly, exogenous NR is a salvageable NAD^+ precursor, and total cellular NAD^+ level of yeast cells can be augmented with NR in media, as long as NA concentrations are restricted to low μM or zero in media. Under these conditions, NR can increase silencing and provide life span increase in a Sir2-dependent manner.[153] Interestingly, it is found that *ΔSir2* strains enhance NR release from yeast cells, which suggests an active role of Sir2 in NR/NAD^+ metabolism.[156]

The yeast system has been remarkably instructive for obtaining a broader and well-studied understanding of how NAD^+ metabolism couples to sirtuin activity. Both major aspects of this regulation, NAD^+ activation and NAM inhibition, appear to be conserved in mammalian systems. In addition, NAD^+ enhancing effects of NR in yeast appear to be conserved to some extent in mammalian cells and tissues, although some important differences have been determined. We discuss NAM and NR effects in mammalian systems in the following sections.

6.4 NAM Metabolite Effects on Mammalian Sirtuin Enzymes

NAD^+ is converted to NAM and O-acetyl-ADP-ribose during sirtuin-mediated deacetylation of target Lys residues, and NAM is a common product of all NAD^+ consumers. NAM is a potent inhibitor of sirtuins[150] that binds to the C-pocket of sirtuins and acts by base exchange to inhibit sirtuin activity.[149] NAM is probably the most universal and biologically relevant sirtuin inhibitor. Recent studies have indicated a link between NAM and its methylated form 1-methylnicotinamide (MNA) and health. In the nematode, both NAM and MNA increase longevity, and overexpression of the nematode gene required for formation of MNA, anmt-1, which is a homolog of the mammalian NAM N-methyltransferase (NNMT), also increases lifespan via a sirtuin-independent pathway.[157] In mammals, knockdown of NNMT protects against diet-induced obesity,[158] although a more recent study found that liver NNMT is increased by calorie restriction and improves glucose and cholesterol metabolism.[159] Physiologic NAM concentrations in brain may be as high as 100 μM, suggesting NAM normally has potent inhibitory effects on sirtuins in mammals.

7. NAD^+ AUGMENTATION AND PROOFS OF CONCEPT FOR THERAPEUTIC INTERVENTIONS

The collective data in yeast and in mammalian cells strongly supports the concept that NAD^+ regulates sirtuin activity. Potent biological effects of increased SIRT1 and SIRT3 activities, such as protection from metabolic syndrome, increased mitochondrial biogenesis, increased ROS detoxification, and neuroprotection invite consideration of NAD^+ augmentation as a means to achieve therapeutic effects if proof of concept studies can be completed and demonstrated. This has largely been the case, where different approaches, such a genetic strategy PARP1 or CD38 deletion, has provided improved activities of sirtuins, such as SIRT1 in particular. A more practical and translational approach to exploit for therapy would be to identify small molecules that either convert to NAD^+, or that inhibit key NAD^+ consuming enzymes leading to phenocopies of PARP1 or CD38 deletion.

Most of the success in obtaining proof of concept for a pharmacologic strategy has been obtained by use of NAD^+ precursors, notably NMN and NR. Boosting NAD^+ levels is largely beneficial[16,123,160] and modulates cellular senescence and ageing. Thus, modulation of NAD^+ production is anticipated to prolong both health span and life span.[14,161] Importantly, NAD^+ precursors provide protection and augmentation to NAD^+ pools without having to sacrifice a key cellular enzymatic function, such as CD38 or PARP1. Elevated NAD^+ levels can also potentially activate all

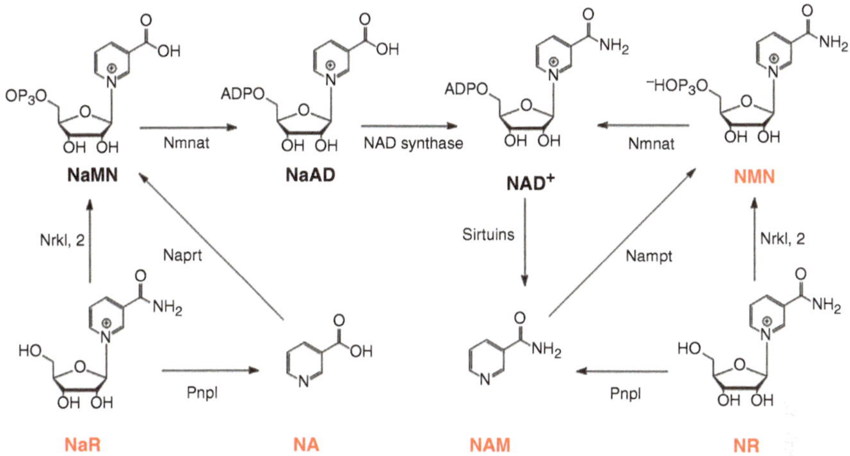

Fig. 4 Pathways of NAD+ biosynthesis in mammalian cells. NAD+ precursors *(red)* can be used to enhance NAD+ concentrations. *NA*, Nicotinic acid; *NaAD*, nicotinic acid adenine dinucleotide; *NAM*, nicotinamide; *NaMN*, nicotinic acid mononucleotide; *Nampt*, nicotinamide phosphoribosyltransferase; *NMN*, nicotinamide mononucleotide; *NR*, nicotinamide riboside.

seven sirtuins. NR, NMN, and NaR precursors (NaR, NA riboside, is largely unstudied) bypass the normal metabolic pathways of NAM and NA (Fig. 4) The vitamin B3 nucleosides NR and NaR can be imported into cells and are thought to be either phosphorylated by Nrk1 or Nrk2 to achieve the respective monophosphates NaMN or NMN. Alternatively, it has been proposed that NMN is directly imported into cells intact. The mononucleotides are further adenylated by nicotinamide mononucleotide adenlyltransferase (NMNAT) as already described (Fig. 4). Available data indicates that NR in particular is a superior NAD+ augmenting agents, leading to two- to three-fold increases in NAD+ in cell culture[162] and significant increases in mammalian tissues, such as liver,[50] muscle,[50] and brain.[163]

For proof of concept pharmacology studies, the starting points and frames of reference are provided by the free base precursors NA and NAM. Importantly, NA has been widely used for treatment of hyperlipidemias and cardiovascular diseases,[164] although it can cause uncomfortable flushing with even intermediate doses. Nevertheless, recent studies indicate that NA can increase NAD+ concentrations in cell culture and in mice tissues.[165,166] Moreover, a new study reports NA improves genome stability.[167] NAM is the most common precursor form of NAD+ in mammalian cells and tissues.

NAM exhibits variable effects on NAD^+ concentrations in different tissues. Liver rapidly converts NAM into increased NAD^+, whereas in other tissues NAM has limited NAD^+ enhancing effects.[168] NAM is typically not limiting for NAD^+ synthesis because of a very low K_m for the enzyme Nampt. NAM also inhibits sirtuins, making it seemingly less useful as an NAD^+ precursor for enhancing sirtuin activity. While some of the data for NA and NAM are quite positive, both NR and NMN appear to exert more potent effects on NAD^+ synthesis, suggesting greater potential for these newer NAD^+ precursors in some applications.

NR is present in cow's milk, thus constituting a dietary source of Vitamin B3 for NAD^+ biosynthesis[169] via two pathways that consist of the Nrk1(2)-dependent route and Pnp-mediated catabolism. NR can strongly promote the biosynthesis of NAD^+ in mammalian cells and tissues.[50,130,160,170] By this means, NR increases the catalytic activity of NAD^+-dependent sirtuin activities.[130,153,171] The Sauve laboratory[162] has developed a synthetic route to prepare NR in amounts suitable for cell studies and animal feeding experiments. We have demonstrated that NR is a potent stimulator of NAD^+ production in various mammalian cell lines from 170% to 270% of controls. NR prevents high fat diet-induced obesity[50] and protects mice from noise-induced hearing loss,[130] DNA damage,[172] and can improve lifespan in an animal model of heart failure.[173] NR treatment in a mouse model of Alzheimer disease, significantly increased the NAD^+ level in the cerebral cortex and improved cognitive function.[163] NR supplementation in old mice (24 months of age) resulted in a significant increase (~40%) in additional lifespan (as measured at the time treatment started).[123] Importantly, a clinical study was published recently found NR is safe at dosages of up to 1000 mg/kg taken for months.[174] These findings suggest that NR can serve as a potent agent for the treatment of aging and aging-related metabolic disorders.

NaR, synthesized from nicotinic acid riboside ethyl ester,[162] has also been identified as a NAD^+ precursor in yeast and mammalian cells, through both Nrk-dependent and Nrk-independent pathway to produce NAD^+.[175,176] It has been reported that NaR treatment increases intracellular NAD^+ concentrations in three different mammalian cell lines by 150%–190% of controls,[162] although in yeast NAD^+ precursor activity of NaR depends upon degradation.[175] NaR has the potential to be important mammalian nutritional precursor, and could enhance NAD^+-dependent benefits provided by sirtuins, although its effects still need to be determined.

NMN can also serve as an NAD^+ precursor, particularly because it has been observed to attenuate age-associated physiological problems and other

Table 2 Studies of NAD+ Precursors on Physiological Effects and Their Effects on Health.

Precursors	Pathways	Sirtuin Isotypes	Physiological Effects
NA	Naprt1	Sir2, SIRT1	Reduce lipids,[164] increase HDL level,[183] genomic stability[167]
NAM	Nampt	Inhibition sirtuins	Treatment of pellagra,[184,185] prevention of vascular injury[186]
NR	Nrk1,2/Pnp1/Urh1/Meu1	Sir2, SIRT1, SIRT3	Longevity,[123] improve muscle function,[160] prevent high fat diet-induced obesity,[50] prevent DNA damage,[172] prevent noise-induced hearing loss,[130] prevent cardiac injury[173]
NaR	Nrk1,2/Pnp1/Urh1/Meu1	SIRT1	Increase NAD levels[175,176]
NMN	Nmnat1–3	SIRT1, SIRT3	Longevity,[177] attenuate brain injury,[187] reverse Alzheimer disease,[188] neuroprotection[182]

NA, Nicotinic acid; NAM, nicotinamide; Nampt, nicotinamide phosphoribosyltransferase; NMN, nicotinamide mononucleotide; NR, nicotinamide riboside; Sir2, silent information regulator 2.

disease conditions in mice.[49,177,178] The effect of NMN as an NAD+ raising agent requires more data, although Revollo *et al.* found that it could increase NAD+ contents by 30% in isolated islets in mice.[179] This suggests that NMN could be an effective precursor of NAD+ in vivo. It has been shown that glucose-stimulated insulin secretion can be corrected by administration of NMN in genetic mouse models or aged mice.[179,180] Treatment with NMN protects old mice against diet- and age-induced diabetes[49] and restores skeletal muscle mitochondrial function to youthful levels.[177] NMN also protects the heart from ischemia and reperfusion injury,[181] and ameliorates cognitive impairment and neuronal death in Alzheimer's disease model rats (Table 2).[182]

ACKNOWLEDGMENTS

AAS and Cornell University receive royalties on sales of nicotinamide riboside from Chromadex Inc, and AAS is a cofounder and consultant of Metro Biotech LLC. The work was partially supported by NIH grant GM106072 and a NY Spinal Cord Injury Board contract to AAS.

REFERENCES

1. Berger F, Ramirez-Hernandez MH, Ziegler M. The new life of a centenarian: signalling functions of NAD(P). *Trends Biochem Sci.* 2004;29:111–118.
2. Imai S, Armstrong CM, Kaeberlein M, Guarente L. Transcriptional silencing and longevity protein Sir2 is an NAD-dependent histone deacetylase. *Nature.* 2000;403:795–800.
3. Haigis MC, Sinclair DA. Mammalian sirtuins: biological insights and disease relevance. *Annu Rev Pathol.* 2010;5:253–295.
4. Houtkooper RH, Pirinen E, Auwerx J. Sirtuins as regulators of metabolism and healthspan. *Nat Rev Mol Cell Biol.* 2012;13:225–238.
5. Schreiber V, Dantzer F, Ame JC, de Murcia G. Poly(ADP-ribose): novel functions for an old molecule. *Nat Rev Mol Cell Biol.* 2006;7:517–528.
6. Bai P, Canto C, Oudart H, Brunyanszki A, Cen Y, Thomas C, Yamamoto H, Huber A, Kiss B, Houtkooper RH, Schoonjans K, Schreiber V, Sauve AA, Menissier-de Murcia J, Auwerx J. PARP-1 inhibition increases mitochondrial metabolism through SIRT1 activation. *Cell Metab.* 2011;13:461–468.
7. Bai P, Canto C, Brunyanszki A, Huber A, Szanto M, Cen Y, Yamamoto H, Houten SM, Kiss B, Oudart H, Gergely P, Menissier-de Murcia J, Schreiber V, Sauve AA, Auwerx J. PARP-2 regulates SIRT1 expression and whole-body energy expenditure. *Cell Metab.* 2011;13:450–460.
8. De Flora A, Zocchi E, Guida L, Franco L, Bruzzone S. Autocrine and paracrine calcium signaling by the CD38/NAD$^+$/cyclic ADP-ribose system. *Ann N Y Acad Sci.* 2004;1028:176–191.
9. Chini EN. CD38 as a regulator of cellular NAD: a novel potential pharmacological target for metabolic conditions. *Curr Pharm Des.* 2009;15:57–63.
10. Xu P, Sauve AA. Vitamin B3, the nicotinamide adenine dinucleotides and aging. *Mech Ageing Dev.* 2010;131:287–298.
11. Bonkowski MS, Sinclair DA. Slowing ageing by design: the rise of NAD$^+$ and sirtuin-activating compounds. *Nat Rev Mol Cell Biol.* 2016;17:679–690.
12. Rine J, Herskowitz I. Four genes responsible for a position effect on expression from HML and HMR in *Saccharomyces cerevisiae*. *Genetics.* 1987;116:9–22.
13. Frye RA. Phylogenetic classification of prokaryotic and eukaryotic Sir2-like proteins. *Biochem Biophys Res Commun.* 2000;273:793–798.
14. Mouchiroud L, Houtkooper RH, Moullan N, Katsyuba E, Ryu D, Canto C, Mottis A, Jo YS, Viswanathan M, Schoonjans K, Guarente L, Auwerx J. The NAD$^+$/sirtuin pathway modulates longevity through activation of mitochondrial UPR and FOXO signaling. *Cell.* 2013;154:430–441.
15. Canto C, Auwerx J. Targeting sirtuin 1 to improve metabolism: all you need is NAD$^+$? *Pharmacol Rev.* 2012;64:166–187.
16. Cerutti R, Pirinen E, Lamperti C, Marchet S, Sauve AA, Li W, Leoni V, Schon EA, Dantzer F, Auwerx J, Viscomi C, Zeviani M. NAD$^+$-dependent activation of Sirt1 corrects the phenotype in a mouse model of mitochondrial disease. *Cell Metab.* 2014;19:1042–1049.
17. Sultani G, Samsudeen AF, Osborne B, Turner N. NAD$^+$: a key metabolic regulator with great therapeutic potential. *J Neuroendocrinol.* 2017;29(10):1–9.
18. Srivastava S. Emerging therapeutic roles for NAD$^+$ metabolism in mitochondrial and age-related disorders. *Clin Transl Med.* 2016;5:25.
19. Magni G, Orsomando G, Raffelli N, Ruggieri S. Enzymology of mammalian NAD metabolism in health and disease. *Front Biosci.* 2008;13:6135–6154.
20. Meyerhof O. Mechanisms of glycolysis and fermentation. *Can J Med Sci.* 1951;29:63–77.
21. Hershberger KA, Martin AS, Hirschey MD. Role of NAD$^+$ and mitochondrial sirtuins in cardiac and renal diseases. *Nat Rev Nephrol.* 2017;13:213–225.

22. Croteau DL, Fang EF, Nilsen H, Bohr VA. NAD^+ in DNA repair and mitochondrial maintenance. *Cell Cycle.* 2017;16:491–492.
23. Lin SJ, Ford E, Haigis M, Liszt G, Guarente L. Calorie restriction extends yeast life span by lowering the level of NADH. *Genes Dev.* 2004;18:12–16.
24. Titov DV, Cracan V, Goodman RP, Peng J, Grabarek Z, Mootha VK. Complementation of mitochondrial electron transport chain by manipulation of the NAD^+/NADH ratio. *Science.* 2016;352:231–235.
25. Ying W. NAD^+/NADH and $NADP^+$/NADPH in cellular functions and cell death: regulation and biological consequences. *Antioxid Redox Signal.* 2008;10:179–206.
26. Sauve AA, Wolberger C, Schramm VL, Boeke JD. The biochemistry of sirtuins. *Annu Rev Biochem.* 2006;75:435–465.
27. Landry J, Slama JT, Sternglanz R. Role of NAD^+ in the deacetylase activity of the SIR2-like proteins. *Biochem Biophys Res Commun.* 2000;278:685–690.
28. Cen Y, Sauve AA. Transition state of ADP-ribosylation of acetyllysine catalyzed by *Archaeoglobus fulgidus* Sir2 determined by kinetic isotope effects and computational approaches. *J Am Chem Soc.* 2010;132:12286–12298.
29. Canto C, Sauve AA, Bai P. Crosstalk between poly(ADP-ribose) polymerase and sirtuin enzymes. *Mol Aspects Med.* 2013;34:1168–1201.
30. Koch-Nolte F, Haag F, Guse AH, Lund F, Ziegler M. Emerging roles of NAD^+ and its metabolites in cell signaling. *Sci Signal.* 2009;2:mr1.
31. Canto C, Auwerx J. NAD^+ as a signaling molecule modulating metabolism. *Cold Spring Harb Symp Quant Biol.* 2011;76:291–298.
32. Galione A. NAADP, a new intracellular messenger that mobilizes Ca^{2+} from acidic stores. *Biochem Soc Trans.* 2006;34:922–926.
33. Fliegert R, Gasser A, Guse AH. Regulation of calcium signalling by adenine-based second messengers. *Biochem Soc Trans.* 2007;35:109–114.
34. Tong L, Denu JM. Function and metabolism of sirtuin metabolite O-acetyl-ADP-ribose. *Biochim Biophys Acta.* 2010;1804:1617–1625.
35. Rechsteiner M, Hillyard D, Olivera BM. Turnover at nicotinamide adenine dinucleotide in cultures of human cells. *J Cell Physiol.* 1976;88:207–217.
36. Fulco M, Cen Y, Zhao P, Hoffman EP, McBurney MW, Sauve AA, Sartorelli V. Glucose restriction inhibits skeletal myoblast differentiation by activating SIRT1 through AMPK-mediated regulation of Nampt. *Dev Cell.* 2008;14:661–673.
37. Yang H, Yang T, Baur JA, Perez E, Matsui T, Carmona JJ, Lamming DW, Souza-Pinto NC, Bohr VA, Rosenzweig A, de Cabo R, Sauve AA, Sinclair DA. Nutrient-sensitive mitochondrial NAD^+ levels dictate cell survival. *Cell.* 2007;130:1095–1107.
38. Burgos ES, Schramm VL. Weak coupling of ATP hydrolysis to the chemical equilibrium of human nicotinamide phosphoribosyltransferase. *Biochemistry.* 2008;47:11086–11096.
39. Koltai E, Szabo Z, Atalay M, Boldogh I, Naito H, Goto S, Nyakas C, Radak Z. Exercise alters SIRT1, SIRT6, NAD and NAMPT levels in skeletal muscle of aged rats. *Mech Ageing Dev.* 2010;131:21–28.
40. Costford SR, Bajpeyi S, Pasarica M, Albarado DC, Thomas SC, Xie H, Church TS, Jubrias SA, Conley KE, Smith SR. Skeletal muscle NAMPT is induced by exercise in humans. *Am J Physiol Endocrinol Metab.* 2010;298:E117–E126.
41. Yang Y, Sauve AA. NAD^+ metabolism: bioenergetics, signaling and manipulation for therapy. *Biochim Biophys Acta.* 2016;1864:1787–1800.
42. Wood JG, Rogina B, Lavu S, Howitz K, Helfand SL, Tatar M, Sinclair D. Sirtuin activators mimic caloric restriction and delay ageing in metazoans. *Nature.* 2004;430:686–689.
43. Smith BC, Hallows WC, Denu JM. A continuous microplate assay for sirtuins and nicotinamide-producing enzymes. *Anal Biochem.* 2009;394:101–109.

44. Jin L, Galonek H, Israelian K, Choy W, Morrison M, Xia Y, Wang X, Xu Y, Yang Y, Smith JJ, Hoffmann E, Carney DP, Perni RB, Jirousek MR, Bemis JE, Milne JC, Sinclair DA, Westphal CH. Biochemical characterization, localization, and tissue distribution of the longer form of mouse SIRT3. *Protein Sci.* 2009;18:514–525.
45. Nakagawa T, Lomb DJ, Haigis MC, Guarente L. SIRT5 deacetylates carbamoyl phosphate synthetase 1 and regulates the urea cycle. *Cell.* 2009;137:560–570.
46. Feldman JL, Dittenhafer-Reed KE, Kudo N, Thelen JN, Ito A, Yoshida M, Denu JM. Kinetic and structural basis for acyl-group selectivity and NAD^+ dependence in sirtuin-catalyzed deacylation. *Biochemistry.* 2015;54:3037–3050.
47. Canto C, Gerhart-Hines Z, Feige JN, Lagouge M, Noriega L, Milne JC, Elliott PJ, Puigserver P, Auwerx J. AMPK regulates energy expenditure by modulating NAD^+ metabolism and SIRT1 activity. *Nature.* 2009;458:1056–1060.
48. Chen D, Bruno J, Easlon E, Lin SJ, Cheng HL, Alt FW, Guarente L. Tissue-specific regulation of SIRT1 by calorie restriction. *Genes Dev.* 2008;22:1753–1757.
49. Yoshino J, Mills KF, Yoon MJ, Imai S. Nicotinamide mononucleotide, a key NAD^+ intermediate, treats the pathophysiology of diet- and age-induced diabetes in mice. *Cell Metab.* 2011;14:528–536.
50. Canto C, Houtkooper RH, Pirinen E, Youn DY, Oosterveer MH, Cen Y, Fernandez-Marcos PJ, Yamamoto H, Andreux PA, Cettour-Rose P, Gademann K, Rinsch C, Schoonjans K, Sauve AA, Auwerx J. The NAD^+ precursor nicotinamide riboside enhances oxidative metabolism and protects against high-fat diet-induced obesity. *Cell Metab.* 2012;15:838–847.
51. Cai AL, Zipfel GJ, Sheline CT. Zinc neurotoxicity is dependent on intracellular NAD levels and the sirtuin pathway. *Eur J Neurosci.* 2006;24:2169–2176.
52. Michan S, Sinclair D. Sirtuins in mammals: insights into their biological function. *Biochem J.* 2007;404:1–13.
53. Michishita E, Park JY, Burneskis JM, Barrett JC, Horikawa I. Evolutionarily conserved and nonconserved cellular localizations and functions of human SIRT proteins. *Mol Biol Cell.* 2005;16:4623–4635.
54. Onyango P, Celic I, McCaffery JM, Boeke JD, Feinberg AP. SIRT3, a human SIR2 homologue, is an NAD-dependent deacetylase localized to mitochondria. *Proc Natl Acad Sci USA.* 2002;99:13653–13658.
55. Vaziri H, Dessain SK, Ng Eaton E, Imai SI, Frye RA, Pandita TK, Guarente L, Weinberg RA. hSIR2(SIRT1) functions as an NAD-dependent p53 deacetylase. *Cell.* 2001;107:149–159.
56. Pruitt K, Zinn RL, Ohm JE, McGarvey KM, Kang SH, Watkins DN, Herman JG, Baylin SB. Inhibition of SIRT1 reactivates silenced cancer genes without loss of promoter DNA hypermethylation. *PLoS Genet.* 2006;2:e40.
57. Meyer S, Chibly AM, Burd R, Limesand KH. Insulin-like growth factor-1-mediated DNA repair in irradiated salivary glands is sirtuin-1 dependent. *J Dent Res.* 2017;96:225–232.
58. Yeung F, Hoberg JE, Ramsey CS, Keller MD, Jones DR, Frye RA, Mayo MW. Modulation of NF-kappaB-dependent transcription and cell survival by the SIRT1 deacetylase. *EMBO J.* 2004;23:2369–2380.
59. Dai JM, Wang ZY, Sun DC, Lin RX, Wang SQ. SIRT1 interacts with p73 and suppresses p73-dependent transcriptional activity. *J Cell Physiol.* 2007;210:161–166.
60. Rodgers JT, Lerin C, Haas W, Gygi SP, Spiegelman BM, Puigserver P. Nutrient control of glucose homeostasis through a complex of PGC-1alpha and SIRT1. *Nature.* 2005;434:113–118.
61. Wang RH, Lahusen TJ, Chen Q, Xu X, Jenkins LM, Leo E, Fu H, Aladjem M, Pommier Y, Appella E, Deng CX. SIRT1 deacetylates TopBP1 and modulates intra-S-phase checkpoint and DNA replication origin firing. *Int J Biol Sci.* 2014;10:1193–1202.

62. Bordone L, Cohen D, Robinson A, Motta MC, van Veen E, Czopik A, Steele AD, Crowe H, Marmor S, Luo J, Gu W, Guarente L. SIRT1 transgenic mice show phenotypes resembling calorie restriction. *Aging Cell.* 2007;6:759–767.
63. Qiao L, Shao J. SIRT1 regulates adiponectin gene expression through Foxo1-C/enhancer-binding protein alpha transcriptional complex. *J Biol Chem.* 2006;281:39915–39924.
64. Jing E, Gesta S, Kahn CR. SIRT2 regulates adipocyte differentiation through FoxO1 acetylation/deacetylation. *Cell Metab.* 2007;6:105–114.
65. Teng YB, Jing H, Aramsangtienchai P, He B, Khan S, Hu J, Lin H, Hao Q. Efficient demyristoylase activity of SIRT2 revealed by kinetic and structural studies. *Sci Rep.* 2015;5:8529.
66. Serrano L, Martinez-Redondo P, Marazuela-Duque A, Vazquez BN, Dooley SJ, Voigt P, Beck DB, Kane-Goldsmith N, Tong Q, Rabanal RM, Fondevila D, Munoz P, Kruger M, Tischfield JA, Vaquero A. The tumor suppressor SirT2 regulates cell cycle progression and genome stability by modulating the mitotic deposition of H4K20 methylation. *Genes Dev.* 2013;27:639–653.
67. North BJ, Rosenberg MA, Jeganathan KB, Hafner AV, Michan S, Dai J, Baker DJ, Cen Y, Wu LE, Sauve AA, van Deursen JM, Rosenzweig A, Sinclair DA. SIRT2 induces the checkpoint kinase BubR1 to increase lifespan. *EMBO J.* 2014;33:1438–1453.
68. Gal J, Bang Y, Choi HJ. SIRT2 interferes with autophagy-mediated degradation of protein aggregates in neuronal cells under proteasome inhibition. *Neurochem Int.* 2012;61:992–1000.
69. Rothgiesser KM, Erener S, Waibel S, Luscher B, Hottiger MO. SIRT2 regulates NF-kappaB dependent gene expression through deacetylation of p65 Lys310. *J Cell Sci.* 2010;123:4251–4258.
70. Eskandarian HA, Impens F, Nahori MA, Soubigou G, Coppee JY, Cossart P, Hamon MA. A role for SIRT2-dependent histone H3K18 deacetylation in bacterial infection. *Science.* 2013;341:1238858.
71. Tao R, Coleman MC, Pennington JD, Ozden O, Park SH, Jiang H, Kim HS, Flynn CR, Hill S, Hayes McDonald W, Olivier AK, Spitz DR, Gius D. Sirt3-mediated deacetylation of evolutionarily conserved lysine 122 regulates MnSOD activity in response to stress. *Mol Cell.* 2010;40:893–904.
72. Bao X, Wang Y, Li X, Li XM, Liu Z, Yang T, Wong CF, Zhang J, Hao Q, Li XD. Identification of "erasers" for lysine crotonylated histone marks using a chemical proteomics approach. *Elife.* 2014;3:1–18.
73. Qiu X, Brown K, Hirschey MD, Verdin E, Chen D. Calorie restriction reduces oxidative stress by SIRT3-mediated SOD2 activation. *Cell Metab.* 2010;12:662–667.
74. Schlicker C, Gertz M, Papatheodorou P, Kachholz B, Becker CF, Steegborn C. Substrates and regulation mechanisms for the human mitochondrial sirtuins Sirt3 and Sirt5. *J Mol Biol.* 2008;382:790–801.
75. Hirschey MD, Shimazu T, Goetzman E, Jing E, Schwer B, Lombard DB, Grueter CA, Harris C, Biddinger S, Ilkayeva OR, Stevens RD, Li Y, Saha AK, Ruderman NB, Bain JR, Newgard CB, Farese Jr RV, Alt FW, Kahn CR, Verdin E. SIRT3 regulates mitochondrial fatty-acid oxidation by reversible enzyme deacetylation. *Nature.* 2010;464:121–125.
76. Kim HS, Patel K, Muldoon-Jacobs K, Bisht KS, Aykin-Burns N, Pennington JD, van der Meer R, Nguyen P, Savage J, Owens KM, Vassilopoulos A, Ozden O, Park SH, Singh KK, Abdulkadir SA, Spitz DR, Deng CX, Gius D. SIRT3 is a mitochondria-localized tumor suppressor required for maintenance of mitochondrial integrity and metabolism during stress. *Cancer Cell.* 2010;17:41–52.
77. Bellizzi D, Rose G, Cavalcante P, Covello G, Dato S, De Rango F, Greco V, Maggiolini M, Feraco E, Mari V, Franceschi C, Passarino G, De Benedictis G. A novel VNTR

enhancer within the SIRT3 gene, a human homologue of SIR2, is associated with survival at oldest ages. *Genomics*. 2005;85:258–263.
78. Ahuja N, Schwer B, Carobbio S, Waltregny D, North BJ, Castronovo V, Maechler P, Verdin E. Regulation of insulin secretion by SIRT4, a mitochondrial ADP-ribosyltransferase. *J Biol Chem*. 2007;282:33583–33592.
79. Rauh D, Fischer F, Gertz M, Lakshminarasimhan M, Bergbrede T, Aladini F, Kambach C, Becker CF, Zerweck J, Schutkowski M, Steegborn C. An acetylome peptide microarray reveals specificities and deacetylation substrates for all human sirtuin isoforms. *Nat Commun*. 2013;4:2327.
80. Jeong SM, Xiao C, Finley LW, Lahusen T, Souza AL, Pierce K, Li YH, Wang X, Laurent G, German NJ, Xu X, Li C, Wang RH, Lee J, Csibi A, Cerione R, Blenis J, Clish CB, Kimmelman A, Deng CX, Haigis MC. SIRT4 has tumor-suppressive activity and regulates the cellular metabolic response to DNA damage by inhibiting mitochondrial glutamine metabolism. *Cancer Cell*. 2013;23:450–463.
81. Miyo M, Yamamoto H, Konno M, Colvin H, Nishida N, Koseki J, Kawamoto K, Ogawa H, Hamabe A, Uemura M, Nishimura J, Hata T, Takemasa I, Mizushima T, Doki Y, Mori M, Ishii H. Tumour-suppressive function of SIRT4 in human colorectal cancer. *Br J Cancer*. 2015;113:492–499.
82. Csibi A, Fendt SM, Li C, Poulogiannis G, Choo AY, Chapski DJ, Jeong SM, Dempsey JM, Parkhitko A, Morrison T, Henske EP, Haigis MC, Cantley LC, Stephanopoulos G, Yu J, Blenis J. The mTORC1 pathway stimulates glutamine metabolism and cell proliferation by repressing SIRT4. *Cell*. 2013;153:840–854.
83. Laurent G, de Boer VC, Finley LW, Sweeney M, Lu H, Schug TT, Cen Y, Jeong SM, Li X, Sauve AA, Haigis MC. SIRT4 represses peroxisome proliferator-activated receptor alpha activity to suppress hepatic fat oxidation. *Mol Cell Biol*. 2013;33:4552–4561.
84. Haigis MC, Mostoslavsky R, Haigis KM, Fahie K, Christodoulou DC, Murphy AJ, Valenzuela DM, Yancopoulos GD, Karow M, Blander G, Wolberger C, Prolla TA, Weindruch R, Alt FW, Guarente L. SIRT4 inhibits glutamate dehydrogenase and opposes the effects of calorie restriction in pancreatic beta cells. *Cell*. 2006;126:941–954.
85. Du J, Zhou Y, Su X, Yu JJ, Khan S, Jiang H, Kim J, Woo J, Kim JH, Choi BH, He B, Chen W, Zhang S, Cerione RA, Auwerx J, Hao Q, Lin H. Sirt5 is a NAD-dependent protein lysine demalonylase and desuccinylase. *Science*. 2011;334:806–809.
86. Zhang Y, Bharathi SS, Rardin MJ, Uppala R, Verdin E, Gibson BW, Goetzman ES. SIRT3 and SIRT5 regulate the enzyme activity and cardiolipin binding of very long-chain acyl-CoA dehydrogenase. *PLoS One*. 2015;10:e0122297.
87. Rardin MJ, He W, Nishida Y, Newman JC, Carrico C, Danielson SR, Guo A, Gut P, Sahu AK, Li B, Uppala R, Fitch M, Riiff T, Zhu L, Zhou J, Mulhern D, Stevens RD, Ilkayeva OR, Newgard CB, Jacobson MP, Hellerstein M, Goetzman ES, Gibson BW, Verdin E. SIRT5 regulates the mitochondrial lysine succinylome and metabolic networks. *Cell Metab*. 2013;18:920–933.
88. Nakamura Y, Ogura M, Ogura K, Tanaka D, Inagaki N. SIRT5 deacetylates and activates urate oxidase in liver mitochondria of mice. *FEBS Lett*. 2012;586:4076–4081.
89. Feldman JL, Baeza J, Denu JM. Activation of the protein deacetylase SIRT6 by long-chain fatty acids and widespread deacylation by mammalian sirtuins. *J Biol Chem*. 2013;288:31350–31356.
90. Liszt G, Ford E, Kurtev M, Guarente L. Mouse Sir2 homolog SIRT6 is a nuclear ADP-ribosyltransferase. *J Biol Chem*. 2005;280:21313–21320.
91. Mao Z, Hine C, Tian X, Van Meter M, Au M, Vaidya A, Seluanov A, Gorbunova V. SIRT6 promotes DNA repair under stress by activating PARP1. *Science*. 2011;332:1443–1446.

92. Jiang H, Khan S, Wang Y, Charron G, He B, Sebastian C, Du J, Kim R, Ge E, Mostoslavsky R, Hang HC, Hao Q, Lin H. SIRT6 regulates TNF-alpha secretion through hydrolysis of long-chain fatty acyl lysine. *Nature.* 2013;496:110–113.
93. Mostoslavsky R, Chua KF, Lombard DB, Pang WW, Fischer MR, Gellon L, Liu P, Mostoslavsky G, Franco S, Murphy MM, Mills KD, Patel P, Hsu JT, Hong AL, Ford E, Cheng HL, Kennedy C, Nunez N, Bronson R, Frendewey D, Auerbach W, Valenzuela D, Karow M, Hottiger MO, Hursting S, Barrett JC, Guarente L, Mulligan R, Demple B, Yancopoulos GD, Alt FW. Genomic instability and aging-like phenotype in the absence of mammalian SIRT6. *Cell.* 2006;124:315–329.
94. Chen S, Seiler J, Santiago-Reichelt M, Felbel K, Grummt I, Voit R. Repression of RNA polymerase I upon stress is caused by inhibition of RNA-dependent deacetylation of PAF53 by SIRT7. *Mol Cell.* 2013;52:303–313.
95. Li L, Shi L, Yang S, Yan R, Zhang D, Yang J, He L, Li W, Yi X, Sun L, Liang J, Cheng Z, Shi L, Shang Y, Yu W. SIRT7 is a histone desuccinylase that functionally links to chromatin compaction and genome stability. *Nat Commun.* 2016;7:12235.
96. Barber MF, Michishita-Kioi E, Xi Y, Tasselli L, Kioi M, Moqtaderi Z, Tennen RI, Paredes S, Young NL, Chen K, Struhl K, Garcia BA, Gozani O, Li W, Chua KF. SIRT7 links H3K18 deacetylation to maintenance of oncogenic transformation. *Nature.* 2012;487:114–118.
97. Kim JK, Noh JH, Jung KH, Eun JW, Bae HJ, Kim MG, Chang YG, Shen Q, Park WS, Lee JY, Borlak J, Nam SW. Sirtuin7 oncogenic potential in human hepatocellular carcinoma and its regulation by the tumor suppressors MiR-125a-5p and MiR-125b. *Hepatology.* 2013;57:1055–1067.
98. Tanno M, Kuno A, Yano T, Miura T, Hisahara S, Ishikawa S, Shimamoto K, Horio Y. Induction of manganese superoxide dismutase by nuclear translocation and activation of SIRT1 promotes cell survival in chronic heart failure. *J Biol Chem.* 2010;285:8375–8382.
99. North BJ, Verdin E. Interphase nucleo-cytoplasmic shuttling and localization of SIRT2 during mitosis. *PLoS One.* 2007;2:e784.
100. Imai S, Johnson FB, Marciniak RA, McVey M, Park PU, Guarente L. Sir2: an NAD-dependent histone deacetylase that connects chromatin silencing, metabolism, and aging. *Cold Spring Harb Symp Quant Biol.* 2000;65:297–302.
101. Di Lisa F, Ziegler M. Pathophysiological relevance of mitochondria in NAD^+ metabolism. *FEBS Lett.* 2001;492:4–8.
102. Pirinen E, Canto C, Jo YS, Morato L, Zhang H, Menzies KJ, Williams EG, Mouchiroud L, Moullan N, Hagberg C, Li W, Timmers S, Imhof R, Verbeek J, Pujol A, van Loon B, Viscomi C, Zeviani M, Schrauwen P, Sauve AA, Schoonjans K, Auwerx J. Pharmacological Inhibition of poly(ADP-ribose) polymerases improves fitness and mitochondrial function in skeletal muscle. *Cell Metab.* 2014;19:1034–1041.
103. Fjeld CC, Birdsong WT, Goodman RH. Differential binding of NAD+ and NADH allows the transcriptional corepressor carboxyl-terminal binding protein to serve as a metabolic sensor. *Proc Natl Acad Sci USA.* 2003;100:9202–9207.
104. Zhang T, Berrocal JG, Frizzell KM, Gamble MJ, DuMond ME, Krishnakumar R, Yang T, Sauve AA, Kraus WL. Enzymes in the NAD+ salvage pathway regulate SIRT1 activity at target gene promoters. *J Biol Chem.* 2009;284:20408–20417.
105. Lau C, Niere M, Ziegler M. The NMN/NaMN adenylyltransferase (NMNAT) protein family. *Front Biosci.* 2009;14:410–431.
106. Cambronne XA, Stewart ML, Kim D, Jones-Brunette AM, Morgan RK, Farrens DL, Cohen MS, Goodman RH. Biosensor reveals multiple sources for mitochondrial NAD^+. *Science.* 2016;352:1474–1477.
107. Motta MC, Divecha N, Lemieux M, Kamel C, Chen D, Gu W, Bultsma Y, McBurney M, Guarente L. Mammalian SIRT1 represses forkhead transcription factors. *Cell.* 2004;116:551–563.

108. Cohen HY, Miller C, Bitterman KJ, Wall NR, Hekking B, Kessler B, Howitz KT, Gorospe M, de Cabo R, Sinclair DA. Calorie restriction promotes mammalian cell survival by inducing the SIRT1 deacetylase. *Science.* 2004;305:390–392.
109. Picard F, Kurtev M, Chung N, Topark-Ngarm A, Senawong T, Machado De Oliveira R, Leid M, McBurney MW, Guarente L. Sirt1 promotes fat mobilization in white adipocytes by repressing PPAR-gamma. *Nature.* 2004;429:771–776.
110. Pfluger PT, Herranz D, Velasco-Miguel S, Serrano M, Tschop MH. Sirt1 protects against high-fat diet-induced metabolic damage. *Proc Natl Acad Sci USA.* 2008;105:9793–9798.
111. Fu M, Liu M, Sauve AA, Jiao X, Zhang X, Wu X, Powell MJ, Yang T, Gu W, Avantaggiati ML, Pattabiraman N, Pestell TG, Wang F, Quong AA, Wang C, Pestell RG. Hormonal control of androgen receptor function through SIRT1. *Molecular and cellular biology.* 2006;26:8122–8135.
112. Dominy Jr JE, Lee Y, Gerhart-Hines Z, Puigserver P. Nutrient-dependent regulation of PGC-1alpha's acetylation state and metabolic function through the enzymatic activities of Sirt1/GCN5. *Biochim Biophys Acta.* 2010;1804:1676–1683.
113. Yoon JC, Puigserver P, Chen G, Donovan J, Wu Z, Rhee J, Adelmant G, Stafford J, Kahn CR, Granner DK, Newgard CB, Spiegelman BM. Control of hepatic gluconeogenesis through the transcriptional coactivator PGC-1. *Nature.* 2001;413:131–138.
114. Diani-Moore S, Ram P, Li X, Mondal P, Youn DY, Sauve AA, Rifkind AB. Identification of the aryl hydrocarbon receptor target gene TiPARP as a mediator of suppression of hepatic gluconeogenesis by 2, 3,7,8-tetrachlorodibenzo-p-dioxin and of nicotinamide as a corrective agent for this effect. *J Biol Chem.* 2010;285:38801–38810.
115. Nakahata Y, Kaluzova M, Grimaldi B, Sahar S, Hirayama J, Chen D, Guarente LP, Sassone-Corsi P. The NAD+-dependent deacetylase SIRT1 modulates CLOCK-mediated chromatin remodeling and circadian control. *Cell.* 2008;134:329–340.
116. Nakahata Y, Sahar S, Astarita G, Kaluzova M, Sassone-Corsi P. Circadian control of the NAD^+ salvage pathway by CLOCK-SIRT1. *Science.* 2009;324:654–657.
117. Ramsey KM, Yoshino J, Brace CS, Abrassart D, Kobayashi Y, Marcheva B, Hong HK, Chong JL, Buhr ED, Lee C, Takahashi JS, Imai S, Bass J. Circadian clock feedback cycle through NAMPT-mediated NAD^+ biosynthesis. *Science.* 2009;324:651–654.
118. Zhang T, Berrocal JG, Yao J, DuMond ME, Krishnakumar R, Ruhl DD, Ryu KW, Gamble MJ, Kraus WL. Regulation of poly(ADP-ribose) polymerase-1-dependent gene expression through promoter-directed recruitment of a nuclear NAD^+ synthase. *J Biol Chem.* 2012;287:12405–12416.
119. Gomes AP, Price NL, Ling AJ, Moslehi JJ, Montgomery MK, Rajman L, White JP, Teodoro JS, Wrann CD, Hubbard BP, Mercken EM, Palmeira CM, de Cabo R, Rolo AP, Turner N, Bell EL, Sinclair DA. Declining NAD^+ induces a pseudohypoxic state disrupting nuclear-mitochondrial communication during aging. *Cell.* 2013;155:1624–1638.
120. Schwer B, North BJ, Frye RA, Ott M, Verdin E. The human silent information regulator (Sir)2 homologue hSIRT3 is a mitochondrial nicotinamide adenine dinucleotide-dependent deacetylase. *J Cell Biol.* 2002;158:647–657.
121. Lombard DB, Alt FW, Cheng HL, Bunkenborg J, Streeper RS, Mostoslavsky R, Kim J, Yancopoulos G, Valenzuela D, Murphy A, Yang Y, Chen Y, Hirschey MD, Bronson RT, Haigis M, Guarente LP, Farese Jr RV, Weissman S, Verdin E, Schwer B. Mammalian Sir2 homolog SIRT3 regulates global mitochondrial lysine acetylation. *Mol Cell Biol.* 2007;27:8807–8814.
122. Someya S, Yu W, Hallows WC, Xu J, Vann JM, Leeuwenburgh C, Tanokura M, Denu JM, Prolla TA. Sirt3 mediates reduction of oxidative damage and prevention of age-related hearing loss under caloric restriction. *Cell.* 2010;143:802–812.
123. Zhang H, Ryu D, Wu Y, Gariani K, Wang X, Luan P, D'Amico D, Ropelle ER, Lutolf MP, Aebersold R, Schoonjans K, Menzies KJ, Auwerx J. NAD^+ repletion improves

mitochondrial and stem cell function and enhances life span in mice. *Science.* 2016;352:1436–1443.

124. Camacho-Pereira J, Tarrago MG, Chini CCS, Nin V, Escande C, Warner GM, Puranik AS, Schoon RA, Reid JM, Galina A, Chini EN. CD38 dictates age-related NAD decline and mitochondrial dysfunction through an SIRT3-dependent mechanism. *Cell Metab.* 2016;23:1127–1139.
125. Rangarajan P, Karthikeyan A, Lu J, Ling EA, Dheen ST. Sirtuin 3 regulates Foxo3a-mediated antioxidant pathway in microglia. *Neuroscience.* 2015;311:398–414.
126. Yu W, Dittenhafer-Reed KE, Denu JM. SIRT3 protein deacetylates isocitrate dehydrogenase 2 (IDH2) and regulates mitochondrial redox status. *J Biol Chem.* 2012;287:14078–14086.
127. Hafner AV, Dai J, Gomes AP, Xiao CY, Palmeira CM, Rosenzweig A, Sinclair DA. Regulation of the mPTP by SIRT3-mediated deacetylation of CypD at lysine 166 suppresses age-related cardiac hypertrophy. *Aging.* 2010;2:914–923.
128. Haigis MC, Deng CX, Finley LW, Kim HS, Gius D. SIRT3 is a mitochondrial tumor suppressor: a scientific tale that connects aberrant cellular ROS, the Warburg effect, and carcinogenesis. *Cancer Res.* 2012;72:2468–2472.
129. Finley LW, Haas W, Desquiret-Dumas V, Wallace DC, Procaccio V, Gygi SP, Haigis MC. Succinate dehydrogenase is a direct target of sirtuin 3 deacetylase activity. *PLoS One.* 2011;6:e23295.
130. Brown KD, Maqsood S, Huang JY, Pan Y, Harkcom W, Li W, Sauve A, Verdin E, Jaffrey SR. Activation of SIRT3 by the NAD^+ precursor nicotinamide riboside protects from noise-induced hearing loss. *Cell Metab.* 2014;20:1059–1068.
131. Finley LW, Carracedo A, Lee J, Souza A, Egia A, Zhang J, Teruya-Feldstein J, Moreira PI, Cardoso SM, Clish CB, Pandolfi PP, Haigis MC. SIRT3 opposes reprogramming of cancer cell metabolism through HIF1alpha destabilization. *Cancer Cell.* 2011;19:416–428.
132. Jeong SM, Lee J, Finley LW, Schmidt PJ, Fleming MD, Haigis MC. SIRT3 regulates cellular iron metabolism and cancer growth by repressing iron regulatory protein 1. *Oncogene.* 2015;34:2115–2124.
133. Ivy JM, Klar AJ, Hicks JB. Cloning and characterization of four SIR genes of *Saccharomyces cerevisiae*. *Mol Cell Biol.* 1986;6:688–702.
134. Lin SJ, Defossez PA, Guarente L. Requirement of NAD and SIR2 for life-span extension by calorie restriction in *Saccharomyces cerevisiae*. *Science.* 2000;289:2126–2128.
135. Rogina B, Helfand SL. Sir2 mediates longevity in the fly through a pathway related to calorie restriction. *Proc Natl Acad Sci USA.* 2004;101:15998–16003.
136. Lamming DW, Latorre-Esteves M, Medvedik O, Wong SN, Tsang FA, Wang C, Lin SJ, Sinclair DA. HST2 mediates SIR2-independent life-span extension by calorie restriction. *Science.* 2005;309:1861–1864.
137. Tissenbaum HA, Guarente L. Increased dosage of a sir-2 gene extends lifespan in *Caenorhabditis elegans*. *Nature.* 2001;410:227–230.
138. Brachmann CB, Sherman JM, Devine SE, Cameron EE, Pillus L, Boeke JD. The SIR2 gene family, conserved from bacteria to humans, functions in silencing, cell cycle progression, and chromosome stability. *Genes Dev.* 1995;9:2888–2902.
139. Xie J, Pierce M, Gailus-Durner V, Wagner M, Winter E, Vershon AK. Sum1 and Hst1 repress middle sporulation-specific gene expression during mitosis in *Saccharomyces cerevisiae*. *EMBO J.* 1999;18:6448–6454.
140. Smith JS, Brachmann CB, Celic I, Kenna MA, Muhammad S, Starai VJ, Avalos JL, Escalante-Semerena JC, Grubmeyer C, Wolberger C, Boeke JD. A phylogenetically conserved NAD^+-dependent protein deacetylase activity in the Sir2 protein family. *Proc Natl Acad Sci USA.* 2000;97:6658–6663.
141. Halme A, Bumgarner S, Styles C, Fink GR. Genetic and epigenetic regulation of the FLO gene family generates cell-surface variation in yeast. *Cell.* 2004;116:405–415.

142. Perrod S, Cockell MM, Laroche T, Renauld H, Ducrest AL, Bonnard C, Gasser SM. A cytosolic NAD-dependent deacetylase, Hst2p, can modulate nucleolar and telomeric silencing in yeast. *EMBO J.* 2001;20:197–209.
143. Wilson JM, Le VQ, Zimmerman C, Marmorstein R, Pillus L. Nuclear export modulates the cytoplasmic Sir2 homologue Hst2. *EMBO Rep.* 2006;7:1247–1251.
144. Celic I, Masumoto H, Griffith WP, Meluh P, Cotter RJ, Boeke JD, Verreault A. The sirtuins hst3 and Hst4p preserve genome integrity by controlling histone h3 lysine 56 deacetylation. *Curr Biol.* 2006;16:1280–1289.
145. Panozzo C, Nawara M, Suski C, Kucharczyka R, Skoneczny M, Becam AM, Rytka J, Herbert CJ. Aerobic and anaerobic NAD^+ metabolism in *Saccharomyces cerevisiae. FEBS Lett.* 2002;517:97–102.
146. Preiss J, Handler P. Biosynthesis of diphosphopyridine nucleotide. I. Identification of intermediates. *J Biol Chem.* 1958;233:488–492.
147. Preiss J, Handler P. Biosynthesis of diphosphopyridine nucleotide. II. Enzymatic aspects. *J Biol Chem.* 1958;233:493–500.
148. Gallo CM, Smith Jr DL, Smith JS. Nicotinamide clearance by Pnc1 directly regulates Sir2-mediated silencing and longevity. *Mol Cell Biol.* 2004;24:1301–1312.
149. Bitterman KJ, Anderson RM, Cohen HY, Latorre-Esteves M, Sinclair DA. Inhibition of silencing and accelerated aging by nicotinamide, a putative negative regulator of yeast sir2 and human SIRT1. *J Biol Chem.* 2002;277:45099–45107.
150. Sauve AA, Schramm VL. Sir2 regulation by nicotinamide results from switching between base exchange and deacetylation chemistry. *Biochemistry.* 2003;42:9249–9256.
151. Jackson MD, Schmidt MT, Oppenheimer NJ, Denu JM. Mechanism of nicotinamide inhibition and transglycosidation by Sir2 histone/protein deacetylases. *J Biol Chem.* 2003;278:50985–50998.
152. Tempel W, Rabeh WM, Bogan KL, Belenky P, Wojcik M, Seidle HF, Nedyalkova L, Yang T, Sauve AA, Park HW, Brenner C. Nicotinamide riboside kinase structures reveal new pathways to NAD^+. *PLoS Biol.* 2007;5:e263.
153. Belenky P, Racette FG, Bogan KL, McClure JM, Smith JS, Brenner C. Nicotinamide riboside promotes Sir2 silencing and extends lifespan via Nrk and Urh1/Pnp1/Meu1 pathways to NAD^+. *Cell.* 2007;129:473–484.
154. Sandmeier JJ, Celic I, Boeke JD, Smith JS. Telomeric and rDNA silencing in *Saccharomyces cerevisiae* are dependent on a nuclear NAD^+ salvage pathway. *Genetics.* 2002;160:877–889.
155. Anderson RM, Bitterman KJ, Wood JG, Medvedik O, Cohen H, Lin SS, Manchester JK, Gordon JI, Sinclair DA. Manipulation of a nuclear NAD^+ salvage pathway delays aging without altering steady-state NAD^+ levels. *J Biol Chem.* 2002;277: 18881–18890.
156. Lu SP, Kato M, Lin SJ. Assimilation of endogenous nicotinamide riboside is essential for calorie restriction-mediated life span extension in *Saccharomyces cerevisiae. J Biol Chem.* 2009;284:17110–17119.
157. Moroz N, Carmona JJ, Anderson E, Hart AC, Sinclair DA, Blackwell TK. Dietary restriction involves NAD^+-dependent mechanisms and a shift toward oxidative metabolism. *Aging Cell.* 2014;13:1075–1085.
158. Kraus D, Yang Q, Kong D, Banks AS, Zhang L, Rodgers JT, Pirinen E, Pulinilkunnil TC, Gong F, Wang YC, Cen Y, Sauve AA, Asara JM, Peroni OD, Monia BP, Bhanot S, Alhonen L, Puigserver P, Kahn BB. Nicotinamide *N*-methyltransferase knockdown protects against diet-induced obesity. *Nature.* 2014;508:258–262.
159. Hong S, Moreno-Navarrete JM, Wei X, Kikukawa Y, Tzameli I, Prasad D, Lee Y, Asara JM, Fernandez-Real JM, Maratos-Flier E, Pissios P. Nicotinamide *N*-methyltransferase regulates hepatic nutrient metabolism through Sirt1 protein stabilization. *Nat Med.* 2015;21:887–894.

160. Ryu D, Zhang H, Ropelle ER, Sorrentino V, Mazala DA, Mouchiroud L, Marshall PL, Campbell MD, Ali AS, Knowels GM, Bellemin S, Iyer SR, Wang X, Gariani K, Sauve AA, Canto C, Conley KE, Walter L, Lovering RM, Chin ER, Jasmin BJ, Marcinek DJ, Menzies KJ, Auwerx J. NAD$^+$ repletion improves muscle function in muscular dystrophy and counters global PARylation. *Sci Transl Med.* 2016;8(361):ra139.
161. Braidy N, Guillemin GJ, Mansour H, Chan-Ling T, Poljak A, Grant R. Age related changes in NAD+ metabolism oxidative stress and Sirt1 activity in wistar rats. *PLoS One.* 2011;6:e19194.
162. Yang T, Chan NY, Sauve AA. Syntheses of nicotinamide riboside and derivatives: effective agents for increasing nicotinamide adenine dinucleotide concentrations in mammalian cells. *J Med Chem.* 2007;50:6458–6461.
163. Gong B, Pan Y, Vempati P, Zhao W, Knable L, Ho L, Wang J, Sastre M, Ono K, Sauve AA, Pasinetti GM. Nicotinamide riboside restores cognition through an upregulation of proliferator-activated receptor-gamma coactivator 1alpha regulated beta-secretase 1 degradation and mitochondrial gene expression in Alzheimer's mouse models. *Neurobiol Aging.* 2013;34:1581–1588.
164. Altschul R, Hoffer A, Stephen JD. Influence of nicotinic acid on serum cholesterol in man. *Arch Biochem Biophys.* 1955;54:558–559.
165. Hara N, Yamada K, Shibata T, Osago H, Hashimoto T, Tsuchiya M. Elevation of cellular NAD levels by nicotinic acid and involvement of nicotinic acid phosphoribosyltransferase in human cells. *J Biol Chem.* 2007;282:24574–24582.
166. Jackson TM, Rawling JM, Roebuck BD, Kirkland JB. Large supplements of nicotinic acid and nicotinamide increase tissue NAD$^+$ and poly(ADP-ribose) levels but do not affect diethylnitrosamine-induced altered hepatic foci in Fischer-344 rats. *J Nutr.* 1995;125:1455–1461.
167. Weidele K, Beneke S, Burkle A. The NAD$^+$ precursor nicotinic acid improves genomic integrity in human peripheral blood mononuclear cells after X-irradiation. *DNA Repair.* 2017;52:12–23.
168. Rawling JM, Jackson TM, Driscoll ER, Kirkland JB. Dietary niacin deficiency lowers tissue poly(ADP-ribose) and NAD$^+$ concentrations in Fischer-344 rats. *J Nutr.* 1994;124:1597–1603.
169. Trammell SA, Yu L, Redpath P, Migaud ME, Brenner C. Nicotinamide riboside is a major NAD+ precursor vitamin in cow milk. *J Nutr.* 2016;146:957–963.
170. Gariani K, Menzies KJ, Ryu D, Wegner CJ, Wang X, Ropelle ER, Moullan N, Zhang H, Perino A, Lemos V, Kim B, Park YK, Piersigilli A, Pham TX, Yang Y, Ku CS, Koo SI, Fomitchova A, Canto C, Schoonjans K, Sauve AA, Lee JY, Auwerx J. Eliciting the mitochondrial unfolded protein response by nicotinamide adenine dinucleotide repletion reverses fatty liver disease in mice. *Hepatology.* 2016;63:1190–1204.
171. Scheibye-Knudsen M, Mitchell SJ, Fang EF, Iyama T, Ward T, Wang J, Dunn CA, Singh N, Veith S, Hasan-Olive MM, Mangerich A, Wilson MA, Mattson MP, Bergersen LH, Cogger VC, Warren A, Le Couteur DG, Moaddel R, Wilson DM, 3rd. Croteau DL, de Cabo R, Bohr VA. A high-fat diet and NAD$^+$ activate Sirt1 to rescue premature aging in cockayne syndrome. *Cell Metab.* 2014;20:840–855.
172. Tummala KS, Gomes AL, Yilmaz M, Grana O, Bakiri L, Ruppen I, Ximenez-Embun P, Sheshappanavar V, Rodriguez-Justo M, Pisano DG, Wagner EF, Djouder N. Inhibition of de novo NAD$^+$ synthesis by oncogenic URI causes liver tumorigenesis through DNA damage. *Cancer Cell.* 2014;26:826–839.
173. Xu W, Barrientos T, Mao L, Rockman HA, Sauve AA, Andrews NC. Lethal cardiomyopathy in mice lacking transferrin receptor in the heart. *Cell Rep.* 2015;13:533–545.
174. Conze DB, Crespo-Barreto J, Kruger CL. Safety assessment of nicotinamide riboside, a form of vitamin B3. *Hum Exp Toxicol.* 2016;35:1149–1160.

175. Belenky P, Christensen KC, Gazzaniga F, Pletnev AA, Brenner C. Nicotinamide riboside and nicotinic acid riboside salvage in fungi and mammals. Quantitative basis for Urh1 and purine nucleoside phosphorylase function in NAD^+ metabolism. *J Biol Chem.* 2009;284:158–164.
176. Kulikova V, Shabalin K, Nerinovski K, Dolle C, Niere M, Yakimov A, Redpath P, Khodorkovskiy M, Migaud ME, Ziegler M, Nikiforov A. Generation, release, and uptake of the NAD precursor nicotinic acid riboside by human cells. *J Biol Chem.* 2015;290:27124–27137.
177. Mills KF, Yoshida S, Stein LR, Grozio A, Kubota S, Sasaki Y, Redpath P, Migaud ME, Apte RS, Uchida K, Yoshino J, Imai SI. Long-term administration of nicotinamide mononucleotide mitigates age-associated physiological decline in mice. *Cell Metab.* 2016;24:795–806.
178. de Picciotto NE, Gano LB, Johnson LC, Martens CR, Sindler AL, Mills KF, Imai S, Seals DR. Nicotinamide mononucleotide supplementation reverses vascular dysfunction and oxidative stress with aging in mice. *Aging Cell.* 2016;15:522–530.
179. Revollo JR, Korner A, Mills KF, Satoh A, Wang T, Garten A, Dasgupta B, Sasaki Y, Wolberger C, Townsend RR, Milbrandt J, Kiess W, Imai S. Nampt/PBEF/Visfatin regulates insulin secretion in beta cells as a systemic NAD biosynthetic enzyme. *Cell Metab.* 2007;6:363–375.
180. Ramsey KM, Mills KF, Satoh A, Imai S. Age-associated loss of Sirt1-mediated enhancement of glucose-stimulated insulin secretion in beta cell-specific Sirt1-overexpressing (BESTO) mice. *Aging Cell.* 2008;7:78–88.
181. Yamamoto T, Byun J, Zhai P, Ikeda Y, Oka S, Sadoshima J. Nicotinamide mononucleotide, an intermediate of NAD^+ synthesis, protects the heart from ischemia and reperfusion. *PLoS One.* 2014;9:e98972.
182. Wang X, Hu X, Yang Y, Takata T, Sakurai T. Nicotinamide mononucleotide protects against beta-amyloid oligomer-induced cognitive impairment and neuronal death. *Brain Res.* 2016;1643:1–9.
183. Penberthy WT. Nicotinic acid-mediated activation of both membrane and nuclear receptors towards therapeutic glucocorticoid mimetics for treating multiple sclerosis. *PPAR Res.* 2009;2009:853707.
184. Harris LJ. Nicotinamide and pellagra; a review of the development of knowledge and of the present position. *Practitioner.* 1952;168:57–59.
185. Raghuramulu N, Srikantia SG, Rao BS, Gopalan C. Nicotinamide nucleotides in the erythrocytes of patients suffering from pellagra. *Biochem J.* 1965;96:837–839.
186. Lin SH, Chong ZZ, Maiese K. Nicotinamide: a nutritional supplement that provides protection against neuronal and vascular injury. *J Med Food.* 2001;4:27–38.
187. Wei CC, Kong YY, Li GQ, Guan YF, Wang P, Miao CY. Nicotinamide mononucleotide attenuates brain injury after intracerebral hemorrhage by activating Nrf2/HO-1 signaling pathway. *Sci Rep.* 2017;7:717.
188. Yao Z, Yang W, Gao Z, Jia P. Nicotinamide mononucleotide inhibits JNK activation to reverse Alzheimer disease. *Neurosci Lett.* 2017;647:133–140.

CHAPTER FOUR

Sirtuins as Modifiers of Huntington's Disease (HD) Pathology

Sin Hui Neo*, Bor Luen Tang*,†,1
* Yong Loo Lin School of Medicine, National University of Singapore, Singapore
† NUS Graduate School for Integrative Sciences and Engineering, National University of Singapore, Singapore
1 Correspondence author. E-mail address: bchtbl@nus.edu.sg

Contents

Abstract

Sirtuins and their pharmacological activators/inhibitors have been associated with a range of neuroprotective effects or disease modifying influences in neurological disorders. Huntington's disease (HD) is an autosomal-dominant, progressive neurodegenerative disease characterized by movement disorder, psychiatric symptoms and cognitive decline. The monogenic mutation in HD encodes a variant of the protein Huntingtin (HTT). The disease is a consequence of a CAG repeat extension leading to

Progress in Molecular Biology and Translational Science, Volume 154
ISSN 1877-1173
http://dx.doi.org/10.1016/bs.pmbts.2017.11.013

an abnormally long polyglutamine (Q) stretch at HTT's N-terminus, which likely confers a toxic gain of function to the mutant polypeptide. HD has currently no effective disease-modifying therapy or preventive measures. In the past 2 decades, a sizable body of work on Sirtuins' modification of HD pathology using HD cell and animal models has accumulated. In this chapter, evidence for Sirtuin activities as potential modifiers of HD pathology is reviewed. The conflicting findings of the impacts of mammalian Sirtuin paralogs on HD pathogenesis and disease progression are highlighted. The possible cellular and molecular mechanisms underlying Sirtuin activities in HD are discussed with reference to pathophysiological mechanisms of transcription perturbation, proteostasis, mitochondrial function, and microtubule dynamics. A brief therapeutic perspective on the use of Sirtuin activators and inhibitors is also presented.

1. INTRODUCTION

1.1 Sirtuins, Aging, and the Brain

The Silent information regulator 2 proteins, or Sirtuins (named after yeast Sir2), constitute a family of protein deacetylases designated as the class III histone deacetylase.[1] Distinct from the class I and II Histone deacetylases (HDACs), Sirtuins' activities are dependent on the oxidized form of nicotinamide adenine dinucleotide (NAD^+) as a cofactor.[2,3] There are altogether seven Sirtuin paralogs in the human genome (SIRT1-7),[4] and these are differentially localized in the cell. SIRT1, SIRT6, and SIRT7 are primarily nuclear, while SIRT3, SIRT4, and SIRT5 are largely mitochondrial. SIRT2 is predominantly cytoplasmic and is a major deacetylating enzyme for tubulin.[5] Other than deacetylase activities, some Sirtuin family members also exhibit a longer acyl chain removal activity, or an ADP-ribosyl transferase activity. The founding member of the family, *Saccharomycescerevisiae* Sir2, was first described as part of a transcription repressor complex for transcriptional silencing of the yeast genomic loci, such as the telomeres, mating type loci, and the ribosomal DNA genes.[6,7] Sir2 suppresses the formation of rDNA circles,[8] and this extends the budding yeast's replicative lifespan.[9,10] The Sirtuins have a broad range of target substrates, and these in combination influence a myriad of important processes that include energy metabolism regulation,[11,12] circadian rhythm control,[13] cellular survival and stress responses,[14] as well as cellular/tissue homeostasis and repair.[15–17] Sirtuin activities may thus impact not just cellular survival but also organismal healthspan and lifespan. The notion that some Sirtuin activities, particularly that of human SIRT1, being systemically beneficial to multiple organs in the later years of life, has important translational implications.

NAD is a key metabolite in cellular energy metabolism and in its reduced form (NADH) is the reducing equivalent that drives oxidative phosphorylation. Sirtuin activities are therefore sensitive to cellular redox status, and would likely be high in conditions of nutrient starvation with high NAD +/NADH ratio. Yeast Sir2's replicative lifespan prolonging activity, which was shown to be activated by resveratrol, a stilbenoid found in abundance in the skin of grapes and berries, is thus proposed to be linked to the effect of caloric restriction (CR), a moderate dietary reduction of food intake without causing malnutrition.[18–20] CR has been widely shown to effectively reduce metabolic disease and conferred measurable lifespan extension in multiple metazoan models.[21] The notion that CR acts through the activities of Sirtuins is attractive and has been intensively investigated,[22] but has remained controversial.[23–25] The Sirtuin-CR link and an apparent ease in Sirtuin activation by a natural plant polyphenol product are, however, the major reasons for the keen pharmacological interest in CR mimetics for metabolic disorders and aging.[26–28] Overexpression or activation of Sir2/Sirt1 was indeed shown to increase mean lifespan in *Caenorhabditis elegans* and *Drosophila*[29–31] models. However, more recent works that revisited these models have however suggested that the lifespan extension effect was less apparent than previously reported.[32–34] On the other hand, yeast Sir2 was shown earlier to oppose the extension of chronogical lifespan.[10] Whether Sirtuins could further extend the much longer mean lifespan in higher organisms and humans has been rather unclear.[24,35] Earlier findings with transgenic overexpression or pharmacological activation of Sirt1 in mice showed improved metabolic parameters in aging animals[36–39] without clear increase in mean lifespan. Several more recent reports have, however, provided stronger indications that some Sirtuin paralogs could indeed extend lifespan in mammalian models. Transgenic overexpression of mouse Sirt6, for example, was shown to increase the lifespan of male mice.[40] The levels of the mitotic checkpoint kinase, Budding uninhibited by benzimidazoles related 1 (BubR1), declines in aging and age-related diseases, and Sirt2 overexpression was found to contextually increase the lifespan of a BubR1 hypomorphic mutant.[41] Furthermore, mice with brain-specific transgenic overexpression of Sirt1 (BRASTO mice, with Sirt1 transgene driven by the PrP promoter) were also shown to have an extended lifespan.[42,43] While resveratrol's effect on mice mean lifespan is unclear, a specific Sirt1 activator SRT1720 extends lifespan of mice on a standard diet.[44]

Brain health is pivotal for the overall health in higher organisms, and CNS neuronal activities that regulate peripheral tissue and organ functions decline with age and neurodegeneration.[45] Accumulating research results suggest that Sirtuins, particularly Sirt1, may exert many systemic effects through its activities in the central nervous system (CNS). Sirt1's deacetylation of transcription factors, such as the Forkhead box class O (FOXO) family members,[46,47] TP53,[48,49] and nuclear factor κB (NF-κB)[50,51] modulates key aspects of stress response and survival pathways in CNS neurons. In *C. elegans*, the FOXO orthologue Daf-16 mediates lifespan extension downstream of the Insulin-like growth factor 1 receptor (IGF-1 receptor/Daf-2)[52] mutant in worms. IGF-1 signaling in neurons has a pivotal role in influencing lifespan as restoring Daf-2 expression in *C. elegans* neurons alone could abolish the lifespan extension effect of Daf-2 mutants.[53] Importantly, IGF-1 signaling in neurons in higher organisms may likewise be influenced by Sirtuins through its FOXO substrates. Sirt1 deacetylates and promotes the activity of key metabolic regulator transcription factors, such as Peroxisome proliferator-activated receptor-γ (PPAR-γ) and its transcriptional coactivator PPARγ coactivator-1α (PGC-1α) and Transducer of regulated CREB2 activity 2 (TORC2) in peripheral tissues.[54,55] In higher organisms, metabolic control is centrally exerted through sensing of energy status in the hypothalamic neurons in the brain, and Sirt1 activities in these neurons are critical for this function.[12,56–59] Sirt1's metabolic regulation is linked to its connection with regulators of the circadian rhythm.[60,61] The central circadian timer protein Circadian locomotor output cycles kaput (CLOCK) has histone acetyltransferase properties, and SIRT1 deacetylase activity regulates the former's transcriptional activity.[62]

Neuronal Sirt1 regulates neurite growth and synaptic processes, and Sirt1 is known to play a role in normal cognitive function and synaptic plasticity.[63,64] Sirt1 activity is also known to affect neural progenitor cell fate during development, as it promotes differentiation of NPCs toward the astroglia lineage.[65] A large body of work also attests to Sirt1's[66,67] and Sirt3's[68] general neuroprotective functions as well as the contrasting activity of Sirt2[69] in a number of neurodegenerative disorders. These findings are further elaborated in sections below.

1.2 Sirtuins' Roles in Neuroprotection/Neurodegeneration

For mammals with limited capacity for adult neurogenesis in the CNS, the repertoire of CNS neurons endowed at birth would need to last a lifetime.

These neurons are however lost with time either due to acute injuries or more chronic neurodegenerative diseases. Prominently amongst the Sirtuins, Sirt1 activity is generally known to be neuroprotective, and this could be intuitively inferred from its deacetylation of a key regulator of apoptosis, TP53,[48,70] in various contexts[71–73] as deacetylation of p53 suppresses apoptosis. Sirt1 is also known to directly and indirectly modulate the expression of neuronal survival factors, particularly the Brain-derived neurotrophic factor (BDNF), whose general beneficial effect in neurodegenerative diseases is well known.[74] For acute neuronal injury, both Sirt1 activation[72,75–77] and Sirt1 overexpression[72] were shown to reduce damages in ischemic stroke and attenuated neuronal death in traumatic brain injuries.[78] Sirt1's neuroprotective effect in acute CNS damage may, however, be context-dependent,[66,79] as several reports have also indicated that Sirt1 inhibition, rather than induction, was neuroprotective in some cases.[80,81] Recent reports have also implicated a beneficial role for Sirt3 in ischemic injury.[82,83] On the other hand, Sirt2 is apparently upregulated in the ischemic brain, and Sirt2 knockout or inhibition alleviated damage in experimental stroke models.[84,85]

Human neurodegenerative diseases are largely sporadic with an age-associated onset, and these result in morbidities, such as cognitive impairment/dementia and movement disorders in the elderly population. The most prevailing cause of age-dependent late-onset dementia is Alzheimer's disease (AD).[86–88] AD histopathology is mainly characterized by the presence of extracellular amyloid plaques [containing amyloid beta (Aβ) peptides] and intraneuronal neurofibrillary tangles (NFT, containing hyperphosphorylated tau)[89,90] in postmortem samples. Aβ peptides are generated from the amyloid precursor protein (APP) by sequential proteases action involving β-secretase[91] and γ-secretase.[92] The formation of Aβ aggregates and neuronal toxicity of Aβ oligomers underlies the amyloid cascade hypothesis of AD etiology.[93] On the other hand, tau-induced pathology also plays a critical role in AD etiology.[94] A majority of AD cases are sporadic, but mutations in both APP and Presenilin, the proteolytic component of the γ-secretase complex, increase Aβ production and promote the earlyonset of AD symptoms.[95] Transgenic expressions of these human mutations in mice form the basis of animal AD models. The most thoroughly investigated Sirtuin family member in connection with AD is Sirt1, which levels appear to be reduced in brain of AD patients.[96,97] A large number of studies have documented a beneficial effect of resveratrol and other compounds on AD-associated pathology, and drug treatment were correlated with an elevation in Sirt1 levels and/or activity using neural cells or neurons in culture[51,98–101] or in mouse and rat models.[22,102–107] Sirt1 overexpression also protected against neurodegeneration in transgenic mice expressing p25, which is the Cyclin-dependent kinase 5 (Cdk5) activator causing tauopathy and AD-like pathology.[102] More recent work has also indicated that Sirt6 may be protective against AD. Sirt6 levels are reduced in an AD mouse model and brain samples from AD patients, while Sirt6 overexpression in a hippocampal neural cell line attenuated Aβ42-induced DNA damage.[108]

Mechanistically, Sirt1's beneficial effects in AD likely occur via modulation of Aβ and tau levels and pathology.[109] Prior cleavage of APP by α-secretase precludes subsequent β-secretase processing and consequential Aβ generation.[110] Overexpression of Sirt1 was shown to enhance the activity of α-secretases, such as A Disintegrin and metalloproteinase domain-containing protein 10 (ADAM10).[111] This led to a decrease in Aβ levels, and corresponding increase in the levels of sAPPα, the α-secretase cleavage product.[112,113] ADAM10 could be activated by compounds, such as the quinolinone-derivative Cilostazol, which occurs through upregulation of Sirt1-coupled retinoic acid receptor β (RARβ) activity.[99] Sirt-mediated transcriptional upregulation of α-secretase may thus lower Aβ levels and suppresses AD pathology. Tau acetylation appears to be an early manifestation in AD brains and is etiologically important for tauopathy,[114] and deacetylation of tau could reduce tau-induced memory deficits and hippocampal atrophy, probably by enhancing its degradation.[115] Tau is shown to be deacetylated by Sirt1, and Sirt1 inhibition resulted in a marked increase in tau acetylation, as well as accumulation of pathogenically hyperphosphorylated tau.[116] Sirt1 activity may also regulate Aβ and tau clearance via macroautophagy.[117,118] Extracellular Aβ could be degraded by metalloproteases, such as the insulin-degrading enzyme (IDE) and neprilysin.[119] Hyperinsulinemia that often occurs in type 2 diabetes mellitus (T2DM) may therefore increase Aβ levels through insulin's competition with Aβ for the enzyme.[120,121] Sirt1 activity, which could metabolically attenuate T2DM, may conceivably enhance Aβ clearance. Notably, resveratrol has been shown to elevate neprilysin levels in mice.[122]

The major cause of hypokinetic movement disorder late in life is Parkinson's disease (PD).[123] PD patients suffer from tremor, hypokinetic rigidity, and postural instability,[124] as a result of axonal impairment, loss of synaptic function or connections and eventual demise of dopaminergic neurons in the substantia nigra pars compacta.[125] Primary PD is largely a late-onset and idiopathic disease, while secondary PD could be linked to environmental toxins.[126] Like AD, there is early-onset, familial forms of PD resulting from gene mutations[127,128] encoding proteins, such as α-Synuclein,[129] Parkin,[130] Phosphatase and tensin homolog (PTEN)–induced putative kinase 1 (PINK1),[131] and DJ-1.[132] α-Synuclein is the primary constituent of extracellular aggregates known as Lewy bodies,[133] the histopathological hallmark of PD neurons.[134] A large number of studies have examined the role of Sirtuin using cell and animal models.[135] Interestingly, the mammalian Sirtuin paralogs appear to have conflicting impacts upon PD pathogenesis and disease progression. Sirt1 is widely shown to be neuroprotective of dopaminergic neurons. The neurotoxic effects of drugs, such as 1-methyl-4-phenyl-1,2,3,6-tetrahydropyridine (MPTP) and 6-hydroxydopamine could be attenuated by resveratrol,[136–138] as well as other related compounds[139–141] through Sirt1 activation, whereas Sirt1 silencing or Sirt1 inhibitors attenuated these effects. With a human α-Synuclein transgenic over-expression model in *C. elegans*, the worm *Sir2.1* was identified as a suppressor of inclusion formation.[142] Transgenic over-expression of the Sirt1 substrate PGC-1α in mice promoted resistance of mice dopaminergic neurons to MPTP.[138] Elevation of mitochondrial Sirt3 has also been shown to protect against neuronal death induced by MPTP[143] and rotenone.[144] Sirt3 and Sirt5 deletion in mice significantly increase nigrostriatal neuron death by MPTP.[145,146] On the other hand, Sirt2 activity exacerbated neuronal demise in PD models,[69,147,148] respectively, and Sirt2 knockout mice have reduced MPTP-induced nigrostriatal damage.[149]

A few key mechanisms may underlie the beneficial effect of Sirt1 in PD. Oxidative damage and mitochondrial dysfunction[150–152] are major pathological findings in PD models. Sirt1 could activate the cellular antioxidant pathway in neurons through PGC-1α.[153–156] α-Synuclein aggregate formation is a key histopathological hallmark of PD neurons.[157] and Sirtuins activity could potentially affect α-Synuclein induced toxicity by either reducing aggregate formation, or enhancing aggregate clearance. In this regard, it is worth noting that Sirt1 regulates the activity of Heat shock factor 1 (HSF1)[158,159] and the expression of chaperone proteins such as Hsp70.[160] Sirt1 and Sirt3 are also both known to positively regulate autophagy/mitophagy,[117] which would help in the clearance of toxic α-Synuclein aggregates and damaged mitochondria. On the other hand, a possible reason for Sirt2's negative effect on PD neurons could be that its deacetylation of α-tubulin modifies microtubule stability and dynamics,[161] and in some yet undefined ways perturbs microtubule-based axonal transport and/or clearing of α-Synuclein.[162] Microtubule instability has in fact been postulated to fuel PD progression.[163] Diseased mitochondria with lower ATP output may also impair microtubule-dependent traffic,[164] and improvement of microtubule-mediated traffic with a polymerizing agent was shown to restore autophagy flux and clearance.[165] PD-associated mutants of LRRK2 are known to interact with deacetylated microtubules and inhibit axonal transport in neurons in a *Drosophila* model, and silencing of Sirt2 rescued both axonal transport and locomotor behavior.[166] These findings may potentially explain why Sirt2's tubulin deacetylation activity could negatively impact neuronal survival in PD.

Other than AD and PD, Sirtuins have also been implicated as potential disease modifiers in other, less common neurodegenerative diseases. Amyotrophic lateral sclerosis (ALS) is a rapidly progressive and fatal motor neuron disease.[167] While most ALS cases are sporadic, about 5%–10% of cases are hereditary in nature (familial ALS), and there are now a large number of implicated gene mutations.[168] The first gene reported for fALS is the Cu-Zn superoxide dismutase gene *SOD1*. Transgenic overexpression of SOD1 mutants, such as SOD1G93A, recapitulates key neurodegenerative features of ALS,[169] and is widely used in ALS cellular and animal models with regards to drug testing. Early work has shown that resveratrol and Sirt1 overexpression significantly attenuated SOD1G93A-mediated neurotoxicity in transfected primary neurons.[102] Exogenous Sirt3 overexpression also promoted neuronal survival and counters mitochondrial fragmentation and impairment of mitochondrial transport in culture rat spinal cord motor neurons induced by SOD1G93A.[170] Astrocytes are critical nonneuronal determinants of ALS progression and motor neuron demise,[171] and overexpression of Sirt3 attenuated the motor neuron toxicity conferred by SOD1G93A bearing astrocytes.[172] Peritoneal injection of resveratrol into SOD1G93A mice delayed disease onset and extended survival,[173] and this effect is associated with deacetylation of HSF1 and upregulation of Hsp25 and Hsp70 in the spinal cord. A resveratrol enriched diet could also delay disease onset and improve function and survival of spinal motor neurons in SOD1G93A mice.[174] Transgenic elevation of CNS Sirt1 expression was shown to extend the lifespan and slowed disease progression in a SOD1G93A line with lower levels of the mutant protein,[175] and this is again associated with deacetylation of HSF1 and increase Hsp70 expression. Overexpression of the Sirt1 substrate PGC-1α also significantly improved the motor function and survival of SOD1G93A mice.[176] In SOD1 mutant ALS models, the beneficial effect of Sirt1 and Sirt3 shown could have also occurred via countering mitochondrial dysfunction and oxidative stress, as well as autophagic clearance of toxic mutant SOD1 proteins or their aggregates.[177] The latter mechanism also likely underlies Sirt1's beneficial effect in prion disease models.[178,179]

1.3 Huntington's Disease (HD) and Mutant Huntingtin (HTT)

Huntington's disease (HD) is a progressive, fatal hereditary autosomal dominant neurodegenerative disorder with multiple neurological manifestations.[180,181] The prevalence of the disease in the Caucasian population is 4/5–10 per 100,000.[182] HD has a mean onset age of 40, but symptoms could well appear earlier in life. Known also as Huntington's chorea, a typical HD symptom is involuntary jerky movement. However, this may be preceded at the earlier (or prodromal) phase by a general lack of movement coordination and an unsteady gait. Neurological symptoms of HD also include psychiatric disturbances and cognitive decline, which eventually progresses into dementia. Patients' condition gradually worsens over the course of 15–20 years toward their eventual death. The genetic basis of HD is a CAG trinucleotide repeat expansion within exon 1 of the huntingtin gene (*HTT*)[183] located on human chromosome 4 at 4p16.3, which translates into a polyglutamine (polyQ) tract in the N-terminus of the huntingtin protein (HTT),[184] and HD is a prototypical polyQ disease.[185] Disease confirmation is based on genetic testing, with 40 or more CAG repeats considered as a full-penetrance allele and thus a positive test result. Clinical assessment based on the unified Huntington's disease rating scale[186] can determine disease onset and monitor progression. HD neuropathology stems from the death of striatal neurons, with prominent loss of GABAergic medium spiny projection neurons projecting to the globus pallidus and the substantia nigra, and neurons that project from the cortex to the striatum. Reduction of the striatum volume and thinning of the cortex could therefore be detected by CT/MRI in genetically diagnosed individuals before the appearance of the symptoms.[187] Prominent neurocytopathological features include ubiquitin positive cytoplasmic aggregates and intranuclear inclusions consisting of N-terminal mutant HTT fragments. The disease has currently no curative therapy, and therapeutic options available provide only symptomatic management of chorea and psychiatric disorder, and do not modify disease progression.[182,188]

The human *HTT* gene, which is conserved in metazoans[189] has 67 exons, encodes two mRNA transcripts, and gives rise primarily to a ubiquitously expressed large (348 kDa) polypeptide. The pathogenic mutation is in exon 1, while the majority part of the polypeptide contains multiple

HEAT repeats (acronym derived from proteins with similar repeats: *H*untingtin, *E*longation factor 3, Protein phosphatase 2*A*, and *T*arget of rapamycin 1).[190] These consist of antiparallel α-helices clustering into larger α-rod domains separated by disordered regions.[191] The repeats form structures that could acts as scaffolds for both inter- and intramolecular interactions.[192–194] HTT has many interacting partners and partake in a myriad of cellular processes in the nucleus and cytoplasm,[184] which is not limited to neurons. The primary sequence of HTT has a nuclear export signal but no canonical nuclear localization signal, but is nonetheless localized to the nucleus and affects transcription, likely serving as a scaffold for transcriptional complexes through its interaction with multiple transcription factors.[195–197] Outside the nucleus, HTT also has multiple cytoplasmic partners, such as Huntingtin-associated protein-1 (HAP1),[198,199] Huntingtin-interacting protein 1 (HIP1),[200] kinesin[201] and dynein,[202] which facilitate microtubule-based transport and vesicular traffic in the exo/endocytic pathway. These interactions and functions impinge on processes, such as cell division,[203] ciliogenesis,[204,205] and autophagy.[193,194,206] The importance of the gene product is demonstrated by the fact that its targeted deletion result in early embryonic lethality with gastrulation defects,[207–209] and HTT has been also shown to be important during neurogenesis and nervous system development.[203,205,210] Given the potentially universal importance of HTT, the rather brain region specific neuronal death in HD is intriguing and requires some explanation.

The HTT protein also contains multiple PEST sequence and is susceptible to cleavage by a host of proteases[211] induced during pathological processes, such as calpain,[212,213] caspases,[214,215] lysosomal cathepsins,[216] and matrix metalloproteinases.[217] The pathology of mutant HTT has been principally attributed to the polyQ repeat in exon 1,[218] and proteolytic cleavage gives rise to toxic N-terminal fragments capable of oligomerization and aggregation.[219,220] Increase protease activities in HD brains may lead to the cytoplasmic generation and accumulation of HTT N-terminal polyQ containing fragments that exerts transcriptional toxicity when translocated into the nucleus.[221] The rather specific susceptibility of striatal neurons to mutant HTT could be explained by several possibilities. The first is down to striatal neuron's dependence on BDNF from cortical projections, whose expression[222] and axonal transport and delivery to the cortico-striatal synapse is dependent on proper HTT function.[223] In fact, endocytosis, axonal retrograde transport, and signaling from BDNF-bound TrkB receptor-bearing signaling endosome in neurons are also dependent on HTT, and would be disrupted by mutant HTT.[224] The dependence of striatal neurons on the cortically produced and transported BDNF would potentially render these most susceptible to HD. Another possible explanation stems from the finding that the small GTP-binding protein Rhes, whose expression is high in the striatum,[225] binds mutant HTT.[226] Rhes binding induces sumoylation of mutant HTT, which reduce aggregation and increase soluble mutant HTT, leading to heightened toxicity. Ectopic expression of Rhes at the cerebellum, a brain area normally not affected by HD, in a HD mice model (N171-82Q) led to motor deficits and cerebellar lesions with a marked loss of Purkinje neuron layer, even during the supposed asymptomatic early period.[227] Understanding the underlying reason for neuronal type susceptibility in HD may lead to important therapeutic clues.

2. DISEASE MODIFYING ROLES OF SIRTUINS IN HD MODELS

Given Sirtuins' roles and influences in neuronal injury and neurodegenerative diseases, the effects of their pharmacological and genetic manipulations have been examined in HD disease models. The various findings are discussed in this section, and reports linking manipulation of Sirtuin levels or activities in metazoan models of HD is summarized in Table 1.

2.1 SIRT1

The consequences of manipulations of the activities and expressions of Sir2 and its orthologues have been examined in multiple in vitro and in vivo models of HD.[240] In a *S. cerevisiae* model, HTT containing polyQs at varying residue number were expressed.[241] Aggregates can be seen in 72Q and were more numerous in 103Q-expressing cells. PolyQ expression increased oxidative stress, and oxidative stress induced by polyQ or a chemical oxidant increased Sir2 expression. Sir2 deleted cells (Δ*sir2* strain) are more susceptible to polyQ induced stress, and this was obvious even at the early exponential phase. The Sir2 activator isonicotinamide decreased mutant polyQ aggregation and the consequent stress generated, indicating that Sir2 is protective against polyQ induced toxicity in yeast. *C. elegans* has no HTT orthologue. However, earlier work has also examined the role of *C. elegans sir-2.1* in modulating the toxic effect of transgenic mutant HTT in worm.[228] Increased *sir-2.1* expression rescued neuronal dysfunction caused by an exon 1–like N-terminal fragment with 128Q (Nter 128Q), while a null mutation

Table 1 A Summary of Reports Linking Manipulation of Sirtuin Levels or Activities to Neuroprotective/Neurodegenerative Outcomes in Huntington's Disease (HD) Metazoan Models.

HD Model	Sirtuin Activity/Levels Manipulation	Main Findings	References
SIRT1			
Transgenic *C. elegans* with Htt N-terminal fragment (128Q)	Increased dosage of *C. elegans sir-2.1*[a]	Increased dosage of *sir2.1* rescued early neuronal dysfunction	Parker et al.,[228]
HdhQ111 knock-in mice	Resveratrol	Resveratrol rescued mutant polyQ-induced cell death in striatal neurons cultured from HdhQ111 knock-in mice	
D. melanogaster transgenic for human Htt exon 1 protein (Httex1p Q93)	Gene dosage of *Drosophila* dSir2[b]	Flies that are heterozygous for *Drosophila* dSir2 exhibit improved survival and sparing of photoreceptor neurons compared to wild-type	Pallos et al.,[229]
N171-82Q HD mice	Resveratrol	No improment of motor performance, loss of body weight, and survival	Ho et al.,[230]
R6/2 HD mice	Crossings with *Sirt1* brain-specific knockout mice (BSKO, genotype *Sirt1*$^{flox/flox}$; nestin-cre)	Brain-specific knockout of Sirt1 exacerbated brain pathology of HD mice; whereas overexpression of Sirt1 improves survival, neuropathology, and BDNF expression in HD mice	Jeong et al.,[231]
N171-82Q HD mice BACHD HD mice	Crossings with *Sirt1* transgenic driven by a prion protein promoter	SIRT1 overexpression significantly improves motor function, reduces brain atrophy, and attenuates mutant Htt-mediated metabolic abnormalities in both HD mouse models	Jiang et al.[232]
R6/1 HD mice	Nicotinamide (NAM)	NAM treatment improved motor deficits associated with the HD phenotype and increased BDNF levels	Hathorn et al.,[233]

N171-82Q HD mice	SRT2104	SRT2104 penetrated the blood–brain barrier, attenuated brain atrophy, improved motor function, and extended survival in HD mice	Jiang et al.,[234]
D. melanogaster transgenic for human Htt exon 1 protein (Httex1p Q93)	Selisistat	Selective Sirt1 inhibition rescues brain pathology in *Drosophila* and mice HD model	Smith et al.,[235]
R6/2 HD mice			
SIRT2			
D. melanogaster transgenic for human Htt exon 1 protein (Httex1p Q93)	Gene dosage of *Drosophila* dSirt2	Reduction of dSirt2 led to greater survivor of photoreceptor neurons compared to wild-type	Pallos et al.,[229]
D. melanogaster transgenic for human Htt exon 1 protein (Httex1p Q93)	Sirt2 inhibitors AGK2 and AK-1	Sirt2 inhibitors are neuroprotective in multiple HD models apparently through decreasing sterol biosynthesis	Luthi-Carter et al.,[236]
Transgenic *C. elegans* with Htt N-terminal fragment (128Q)			
Htt171-82Q expression in primary striatal neurons			
R6/2 HD mice	Crossing with Sirt2 knockout mice	Sirt2 ablation had no effect on HD pathology and progression	Bobrowska et al.,[237]
R6/2 HD mice	Sirt2 inhibitor AK7	The Sirt2 inhibitor AK7 improved motor function, extended survival, reduced brain atrophy, and aggregated mutant huntingtin	Chopra et al.,[238]
Drosophila and mice brain slice cultures expressing Htt exon 1 with expanded polyQ	Sirt2 inhibitor MIND4	The Sirt2 inhibitor MIND4 is neuroprotective in both *Drosophila* rhabdomeres and mice brain slice explant	Quinti et al.,[239]

[a] *C. elegans sir-2.1* has the highest similarity to yeast *Sir2*, as for mammalian *Sirt1*.
[b] *D. melanogaster* dSir2 has the highest similarity to yeast *Sir2*, as for mammalian *Sirt1*.

of the gene exacerbated dysfunction. This rescue effect is mimicked by resveratrol, and is lost when either *sir-2.1* or the worm FOXO orthologue, *daf-16*, is silenced. On the other hand, *Drosophila melanogaster* does have a HTT orthologue with functions mirroring those of mammalian HTT.[194,242] Interestingly and in contrast with the findings in the more naïve models of yeast and worm, neurons of transgenic *Drosophila* expressing a HTT exon 1 fragment with 93Q (Httex1p Q93) survived better in a dSir2 null heterozygous background.[229] Feeding these transgenic flies with general Sirtuin inhibitors, such as sirtinol and nicotinamide also promoted neuronal survival. The finding is corroborated by a more recent study with the Httex1p Q93 transgenic fly model.[235] Transgenic flies heterozygous for dSir2 show improved motor function compared to siblings with wild-type dSir2 levels. The authors also showed that a highly selective Sirt1/dSir2 inhibitor, selisistat, rescued neuronal degeneration, and extended lifespan of Httex1p Q93 transgenic flies. Importantly, the inhibitor's effect is lost in flies homozygous for a null allele of dSir2. dSir2's influence or its modulation mutant HTT toxicity in flies thus appears to differ from that in yeast and worm.

The effects of Sirt1 level/activity manipulations have likewise been investigated in a number of mouse HD models,[243] and these also revealed contrasting results. In a N171-82Q transgenic mouse model of HD,[244] a formulation of resveratrol (SRT501), increased expression of PGC-1α in brown adipose tissues (BAT), but not in the striatum.[230] While some peripheral parameters associated with adipose tissues improved, there was no significant improvement in striatal atrophy, motor performance and overall survival. On the other hand, two back-to-back reports showed that Sirt1 is an important positive modulator against HD pathology. Using the R6/2 HD mouse model,[245] Jeong *et al.* showed that brain-specific knockout of Sirt1 (BSKO, Sirt1$^{flox/flox}$; nestin-cre) exacerbated neuropathological phenotype in HD mice, whereas overexpression of Sirt1 improves neuropathology and survival.[231] The authors also identified a putative mechanism for Sirt1's effect, namely enhanced expression of BDNF through deacetylation of transducers of regulated CREB activity 1 (TORC1). Mutant HTT lowers BDNF expression by interfering with TORC1–CREB interaction, while Sirt1 deacetylation promotes TORC1 dephoshorylation and interaction with CREB. In another study, Jiang *et al.* crossed BRASTO mice with high transgenic Sirt1 levels in the brain with either the N171-82Q HD mice or the BACHD mice harboring full-length human mutant HTT,[246] and showed that overexpression of Sirt1 in the brain delayed onset and slowed progression of motor deficits and improved metabolic parameters.[232] The authors showed that Sirt1 prevents mutant HTT-induced decline in BDNF levels and signaling, and that mutant HTT could interact with SIRT1 and inhibit the latter's deacetylase activity. In more recent work, Jiang and coworkers also found that the Sirt1 activator SRT2104[247] could effectively penetrate the blood–brain barrier (BBB) and improved motor function, as well as survival of N171-82Q HD mice.[234]

A major contradicting result pertaining to Sirt1's role in countering HD pathology came in the form of studies with Sirt1 inhibitors. Hathorn *et al.*[233] showed that nicotinamide, a product of Sirtuin-catalyzed deacetylase reaction and an inhibitor of its enzymatic activity, increased BDNF and PGC-1α levels, as well as improved motor dysfunction in R6/1 HD mice.[218] While these results appear to run counter to that of Sirt1 activators, there are potential caveats. It has been recently argued that while nicotinamide does inhibit Sirtuin activity in vitro, in cells and tissues it is very quickly converted to NAD^+, potentially resulting in a net effect of stimulating Sirtuin activity.[248] However, results from another study with the Sirt1 specific inhibitor selisistat provided more robust evidence that inhibition of Sirt1 activity may be beneficial in HD. Administration of selisistat to R6/2 HD mice alleviated some neurological phenotype and prolonged survival.[235] These conflicting findings may be difficult to reconcile. However, it is possible that both activation and inhibition of SIRT1 alleviates different pathological aspects of HD via different mechanisms. Specifically, activating SIRT1 enhances gene transcription of genes for neuronal survival, whereas inhibiting SIRT1 may enhance the clearance of mutant HTT via the acetylation of crucial amino acid residues in mutant HTT which targets them for autophagy.[249]

2.2 SIRT2

Unlike Sirt1, a negative role for Sirt2 in HD pathology appears somewhat more consistent. Flies heterozygous for dSirt2 have better photoreceptor neuron survival with Httex1p Q93 transgenic expression compared to wild-type.[229] It is worth noting, however, that in this fly HD model, a gene dosage reduction in dSir2 also has a similar effect. These results were reproduced by another study with the same fly model.[236] The focus of this later study is however the use of Sirt2 inhibitors AGK2 and AK-1, the latter was previously shown to rescue α-Synuclein mediated neurotoxicity in PD models.[69] These Sirt2 inhibitors also rescued neuronal dysfunction in the *C. elegans* Nter 128Q model, as well as neurotoxicity in rat primary striatal neurons conferred by lentiviral-mediated Htt171-82Q expression.[236] In a follow up study using two HD mouse models, AK7 improved neuropathological phenotypes and extends survival of R6/2 mice and improved motor activity and aggregate formation in the striatum in a 140CAG knock-in mice.[238] Another recent report also showed that a novel thiazole-containing Sirt2 inhibitor, MIND4, was protective of rat corticostriatal slice explants and flies expressing mutant HTT exon 1.[239] However, in one study using the R6/2 model, Sirt2 ablation did not improve HD phenotype, and neither Sirt2 silencing nor knockout changes the oligomeric or aggregate load of mutant HTT.[237] The reason for these discrepancies is unclear.

2.3 SIRT3

A role in mitochondrial Sirt3 in modulating HD pathology was implicated by a study with the stilbenoid *trans*-(−)-ε-Viniferin (viniferin), a novel Sirt3 activator.[250] Viniferin reduced mutant HTT-induced oxidative stress, PGC-1α depletion and mitochondrial dysfunction in immortalized striatal progenitors cells generated from Hdh(Q111) knock-in embryos[251]. Interestingly, the compound also prevented the decline of Sirt3 levels and activity in the HD cell model. The protective effect of viniferin was Sirt3dependent, as this was abolished by Sirt3 silencing.[250]

3. SIRTUINS AS MODIFIERS OF HD—POSSIBLE UNDERLYING MECHANISMS

As summarized in the sections earlier, there appear to be ample evidence for Sirtuins' potential in modifying HD disease phenotype and progression, at least in animal models. How Sirtuins may exert their disease modifying function is a question for which definitive answers are still lacking. The better known possibilities and conceivable mechanisms of action are discussed further below.

3.1 Sirtuin Level Changes in HD and Sirt1–HTT Interaction

All three mammalian Sirtuin paralogs that have been tested in HD models are ubiquitously expressed and found in varying levels in the brain. Sirt1 is abundant in adult rodent and human brains,[63,252,253] and particularly abundant in neurons,[63] but is also expressed in astroglia and oligodendroglia.[97] Brain Sirtuin levels are, as expected, metabolically regulated, and Sirt1, for example, is induced by physical activity[254,255] and CR,[22,42,256] and conversely downregulated by a high-calorie diet.[257,258] Sirt3 levels in brain neurons are likewise increased by exercise.[259] Sirt2 exhibits abundant and preferential splice isoform expression in the brain,[5] particularly in neurons[5,236] and oligodendroglia.[260] Sirtuin levels in the brain changes differentially with age[261]. Sirt1 levels are generally reduced in aged neurons,[256,261] whereas Sirt3 levels showed the opposite trend.[261] Interestingly, a particular splice isoform of Sirt2 accumulates in the aging mouse brain.[5]

Sirt1 levels are generally reduced in neuropathological conditions, such as AD,[96,97,262] and its activity is downregulated in HD models.[263] Sirt3 levels are likewise reduced in cells expressing mutant HTT.[250] Sirt2 levels, however, does not appear to change significantly, at least for the R6/2 mice and striatal neurons expressing mutant HTT.[236] A potentially important, albeit poorly defined, mechanism of Sirt1 in countering the effect of mutant HTT in some of the HD models is an apparent direct interaction between the two.[231,232] Functional activity aside, the overexpressed Sirt1 could therefore potentially bind and reduce the toxicity of nuclear translocated mutant HTT, thus alleviating the effect of the latter on transcriptional inhibition.

3.2 Countering Transcriptional Defects Associated With Mutant HTT

An important aspect of mutant HTT-mediated neurotoxicity is perturbations in nuclear transcription. A particularly critical transcript for neuronal survival is that encoding BDNF, and it is known that wild-type HTT upregulates, while mutant HTT reduces BDNF expression.[222] As discussed in section 2.1, mutant HTT was shown to lower BDNF expression by interfering with TORC1–CREB interaction, while Sirt1 deacetylation promotes TORC1 dephoshorylation and its interaction with CREB.[231] Mutant HTT also represses the expression of PGC-1α,[264,265] which impairs mitochondrial biogenesis resulting in mitochondrial dysfunctions. The role of Sirt1 in modulating other HTT interacting factors, such as the general neural gene suppressor RE1-silencing transcription factor (REST/NRSF)[197] is less clear, although Sirt1 activity has been connected to the transcription factor.[266] Sirt1 could thus functionally compensate for the reduction in BDNF transcription and the repression of PGC-1α (see also Section 3.4) caused by mutant HTT. As mentioned earlier, Sirt1 could also interact directly with mutant HTT, and overexpressing Sirt1 could thus conceivably reduce mutant HTT's transcriptional inhibition via a more physically direct mechanism, namely by physically preventing mutant HTT from disrupting transcriptional complexes.

3.3 Promotion of Proteostasis and Autophagy

Another important aspect of HD pathogenesis is the formation of oligomers from mutant HTT and the accumulation of various proteolytic cleavage products with time. Accumulation of these products, coupled to diminished proteostasis capacity with age,[267] could result in neuronal dysfunction and demise. Proteostasis mechanisms within the cell range from prevention of misfolding and aggregate formation to clearance via the ubiquitin-proteasome pathway and autophagy. Impairment of chaperone activities, aggregate formation and autophagy defects are prominently known in HD.[268–270] A recent discovery has made firm mechanistic connections between polyQ tracts of proteins with autophagy, and explains how autophagy could be impaired in HD.[271] The polyQ domain of wild-type ataxin 3[272] interacts with the autophagy initiator beclin 1, which protects the latter from proteasome-mediated degradation. Soluble proteins with polyQ tracts compete for this ataxin 3–beclin 1 interaction in a length-dependent fashion, thus impairing autophagy. This impairment was indeed observed in cells and mouse models expressing mutant HTT exon 1, as well as cells from HD patients.[271]

Sirt1 is a known regulator of proteostasis as it regulates the activity of Heat shock factor-1 (HSF-1).[273] Deacetylation stabilizes HSF-1's binding to the promoters of heat shock protein genes like *HSP70*, thus enhancing their expression[158]. Over-expression of the chaperones of the heat shock protein family has indeed been shown to be neuroprotective in HD models.[274] Sirt1 also deacetylates some cytoplasmic chaperonins, such as cytoplasmic chaperonin containing TCP-1 (CCT), enhancing their activities.[275,276] Some members of the Sirtuin family are also known modulators of autophagy[117]. Sirt1 could regulate autophagy either by directly deacetylating key autophagy factors like Atg5/7/8,[277] or indirectly via the deacetylation and activity modulation of transcription factors, such as FOXO, TP53, and NF-κB, or via crosstalk with the AMP activated protein kinase (AMPK).[278] Sirt3 has also been implicated in autophagy regulation.[279,280] In neurons, its involvement in autophagy has been shown to be protective against ischemic insult.[83] The role of Sirt2 in autophagy is less clear, but Sirt2 silencing has been shown to increase basal autophagy levels in cancer cells.[281] Taken as a whole, these findings are in line with a largely positive role of Sirt1 and Sirt3, and a negative one for Sirt2, in terms of neuroprotection via proteostatic mechanisms.

3.4 Countering Mitochondrial and Metabolic Dysfunctions

Mitochondrial dysfunction is a prominent feature of common neurodegenerative diseases, such as AD[282] and PD.[150,151] These dysfunctions include changes in mitochondrial dynamics and energy metabolism,[152] both of which predispose neurons to oxidative damage and energy deprivation. In HD, mitochondrial dysfunction could stem from mutant HTT's repression of a key transcription factor for mitochondrial genes, PGC-1α.[264,265] PGC-1α regulates mitochondria biogenesis[283] as well as the induction of several ROS detoxifying enzymes.[284] Sirt3 is an important factor downstream of PGC-1α in terms of mitochondrial function,[285] with its deacetylase activity determining the acetylation status and enzymatic activity of several mitochondrial metabolic enzymes,[286,287] and protecting mitochondria against oxidative damage through FOXO3 activation.[288] Metabolic dysfunction is often a clinical feature of HD, and it is worth noting that early studies have shown that CR confers a beneficial effect in HD mice,[289] thus implicating the involvement of Sirtuins. This is not surprising as PGC-1α is a prominent Sirt1 substrate,[290–292] and its activation by Sirt1 deacetylation has been extensively implicated in metabolic control and mitochondrial biogenesis.[293–295] The Sirt1–PGC-1α connection is also important for defense against mitochondrial oxidative stress.[154] Sirt1 and Sirt3 activity have also been implicated in the regulated disposal of damaged mitochondria through mitophagy.[295] These Sirtuins could thus counter the effect of mutant HTT on mitochondrial function in multiple ways and in tandem.

3.5 Countering Defects in Microtubule-Associated Transport

HTT is involved in the regulation of microtubule-based axonal transport,[296] and mutant HTT causes axonal transport defects.[297,298] Disruption of BDNF transport and deprivation of neurotrophic signaling is indeed a prime mechanism for striatal neuron demise in HD.[223] HTT could be localized to the cilia and mutant HTT could thus also disrupt microtubule-based ciliary transport.[299] Changes in *HTT*'s CAG repeats may perturb the alternative splicing of microtubule binding proteins genes, such as those encoding MAP2 and tau. In fact, tau hyperphosphorylation has been observed in HD models,[300,301] and occurrence of tauopathy in HD[302,303] is an emerging pathological concept that has been previously suspected.[304] Tau is acetylated at K174, and this acetylation retards its turnover.[115] Sirt1 deacetylates tau and enhances its degradation,[116] and so if HD pathology does involve significant tauopathy it would likely be attenuated by Sirt1 activity. More indirectly, dysfunctional mitochondria with lower ATP output may also impair microtubule-dependent traffic, and this could be countered indirectly by Sirt1 activity as discussed in the earlier section.

Unlike Sirt1 and Sirt3, Sirt2 activity has an apparent negative effect in AD,[305] PD,[69,149] and as discussed earlier, also in HD. Although a clear mechanistic understanding for these observations is still lacking, it may conceivably be link to Sirt2's α-tubulin deacetylation activity,[306] which modifies microtubule stability and dynamics,[161] and which may be particularly undesirable for axonal transport that are important for neurotrophic signaling, as well as the clearing of toxic aggregates in a pathological context.

3.6 Suppression of Neuroinflammation

Chronic neuroinflammation is a common characteristic of neurodegenerative disease pathologies.[307] This is likewise the case for HD,[308,309] with mutant HTT promoting microglia activation[310] and generation of neurotoxic astrocytes.[311] Sirtuins, particularly Sirt1, is a suppressor of inflammation in multiple organ systems[312] and the proinflammatory senescence associated secretory phenotype.[313] This activity could be largely down to Sirt1's deacetylation of nuclear factor kappa B (NFκB)'s RelA/p65 subunit at K310, thus inhibiting its transcriptional activity,[50] leading to an attenuation of microglia activation in the brain.[51] Interestingly, Sirt2 also appear to act in the same way in terms of NFκB deacetylation and microglia activation.[314] In ALS, enhancement of the NAD^+ salvage pathway and increasing Sirtuin activity could attenuate neurotoxicity of astrocytes in an ALS mutant SOD1 model.[172] It is conceivable that an increase in Sirtuin activities in the diseased brain could attenuate the generation of neurotoxic reactive A1 type astrocytes driven by proinflammatory microglia that is abundant in brain tissues of AD, PD, ALS, and HD.[315]

4. THERAPEUTIC PERSPECTIVES IN TARGETING SIRTUINS IN HD

HD has currently no curative or disease progression–modifying therapy, and therapeutic options available are limited to those for symptomatic management. The latest major FDA-approved drug in this regard is Deutetrabenazine (trade name Austedo, analog of the first FDA-approved tetrabenazine) for the treatment of chorea, with clear improvement in chorea score recently reported in phase III trials.[316] The more recent major disappointments from clinical testing of possible disease modifiers are coenzyme Q10 and creatine,[317,318] which were hopefuls in terms of improving mitochondrial function and metabolic deficits in HD. Two recent major phase III trials, 2CARE (for coenzyme Q10)[319] and CREST-E (for creatine), were terminated when interim futility analyses indicated very low likelihood of a positive outcome in continuing towards planned endpoints.[320] As manipulation of Sirtuin activities in animal models have been shown to be beneficial to a myriad of aging-related and neurodegenerative disorders, there may be some impetus to extend this to human trials.

Perhaps the simplest way to achieve activation of Sirtuins is the administration of resveratrol and resveratrol-like compounds. Resveratrol likely activates multiple Sirtuins, and although it has low toxicity, it also has limited bioavailability and a number of non-Sirtuin targets.[321] It is the only Sirtuin activator for which a randomized, double-blind placebo-controlled trial on neurodegenerative disease (AD) subjects has been published.[322] This phase II trial for AD patients is insufficiently powered for the detection of clinical outcomes, but did reaffirm that massive doses of resveratrol is well-tolerated for an extended period, and that the compound and its metabolites could indeed penetrate the blood–brain barrier and modestly alter the trajectories of AD biomarkers. The major caveat of using resveratrol in HD is that Sirt2 may also be activated, and in multiple HD models, Sirt2 activation has been shown to be undesirable. Compounds with a higher degree of Sirt1 target specificity, such as SRT2104[241,242] may be more promising in this regard. However, there is currently no HD clinical trial with any specific Sirtuin activators in progress.

On the other hand, the Sirt1 inhibitor Selisistat (EX-527), which has been shown to be protective in fly and mouse HD models[233] has been examined in two human trials. It was shown to be well tolerated in healthy volunteers[323] and was also tested in an exploratory placebo-controlled trial with early stage HD patients.[324] The patient-based trial again demonstrated drug tolerance at the anticipated therapeutic plasma concentration range without adverse motor or cognitive effect, although circulating levels of soluble HTT were not affected. It is unclear if bigger, efficacy-determining trials are being planned with this or other Sirtuin inhibitors.

5. EPILOGUE

In the previous paragraphs, we have outlined how Sirtuins and their activities may act as modifiers of HD pathology. On the whole, Sirt1 and

Sirt3 activity appear to be somewhat neuroprotective, while Sirt2 activity may have the converse effect in diseased conditions. Sirtuins' protective activity could possibly be due to their countering of mutant HTT-induced transcriptional, proteostatic, axonal transport, and mitochondrial, as well as associated metabolic defects. These postulated mechanisms are outlined in Fig. 1. At the moment it is difficult to single out the major mechanisms from lesser ones.

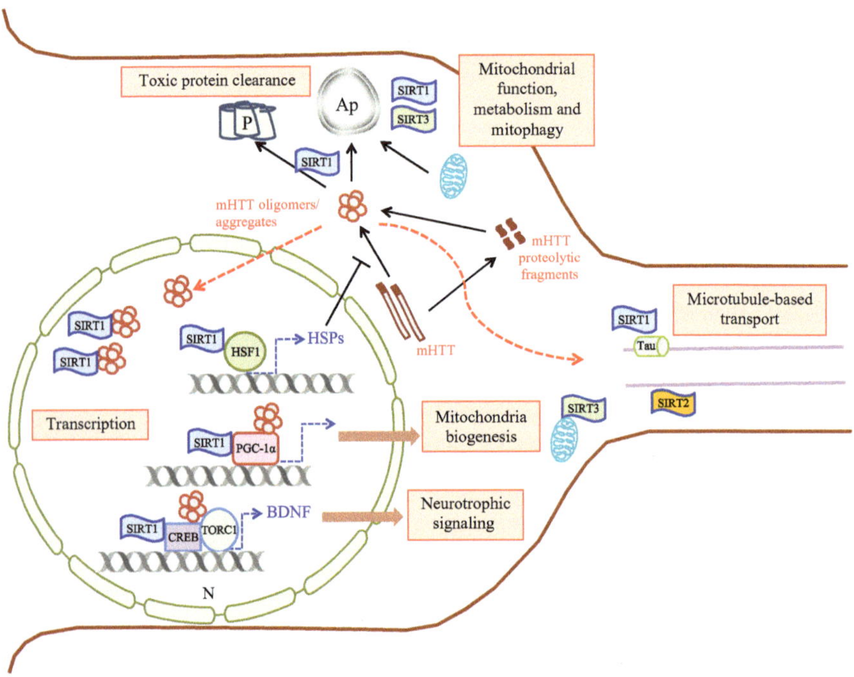

Fig. 1 Major mechanisms by which Sirtuins may affect Huntington's disease (HD) pathology. Nuclear Sirt1 regulates transcription of genes associated with neurotrophic signaling (such as BDNF) to promote neuronal survival, mitochondrial biogenesis, and stress responses (such as production of chaperones by activating HSF1) which would counter generation of mutant HTT (mHTT) oligomers/aggregates. Nuclear Sirt1 counters transcriptional processes disrupted by mutant HTT, such as the interaction between the transcription factors CREB and TORC-1. Cytoplasmic Sirt1 could activate autophagy, facilitate toxic protein clearance, and lower/maintain tau/p-tau levels. Mitochondrial Sirt3 functions along the Sirt1–PGC-1α axis to maintain mitochondrial function and metabolic integrity. Sirt2's deacetylation of α-tubulin may alter microtubule dynamics in unclear ways to enhance HD pathology (see text for details). *Broad arrows* link key transcription processes mediated by nuclear Sirt1 to major protective mechanisms. *Arrows* to and from mHTT oligomers and aggregates indicate their pathological localizations *(red dotted)* and fate *(black solid)*.

A particularly puzzling aspect of the findings thus far pertains to Sirt1, for which either activation or inhibition have been shown to be beneficial. These paradoxical or conflicting results remind us of the complex nature of Sirt1's activity and functions, as well as the limitations of HD animal models as therapeutic testbeds. In most cases the relative activation/inhibition of the Sirtuin paralogs in any of the pharmacological manipulations were unclear, while massive elevations in levels or complete abolishment of any particular Sirtuin via genetic manipulations are unlikely to be optimal. The development of more specific activators/inhibitors, better models, as well as disease biomarkers may help therapeutic advancement. However, the sheer neuropathological complexity of even a monogenic disease like HD would mean that the search for disease modifying therapies in neurodegenerative diseases will remain a very challenging endeavor.

REFERENCES

1. Falkenberg KJ, Johnstone RW. Histone deacetylases and their inhibitors in cancer, neurological diseases and immune disorders. *Nat Rev Drug Discov.* 2014;13(9):673–691.
2. Blander G, Guarente L. The Sir2 family of protein deacetylases. *Annu Rev Biochem.* 2004;73:417–435.
3. Haigis MC, Guarente LP. Mammalian sirtuins: emerging roles in physiology, aging, and calorie restriction. *Genes Dev.* 2006;20(21):2913–2921.
4. Haigis MC, Sinclair DA. Mammalian sirtuins: biological insights and disease relevance. *Annu Rev Pathol.* 2010;5:253–295.
5. Maxwell MM, Tomkinson EM, Nobles J, Wizeman JW, Amore AM, Quinti L, Chopra V, Hersch SM, Kazantsev AG. The Sirtuin 2 microtubule deacetylase is an abundant neuronal protein that accumulates in the aging CNS. *Hum Mol Genet.* 2011;20 (20):3986–3996.
6. Imai S, Armstrong CM, Kaeberlein M, Guarente L. Transcriptional silencing and longevity protein Sir2 is an NAD-dependent histone deacetylase. *Nature.* 2000;403 (6771):795–800.
7. Zhang T, Kraus WL. SIRT1-dependent regulation of chromatin and transcription: linking NAD(+) metabolism and signaling to the control of cellular functions. *Biochim Biophys Acta.* 2010;1804(8):1666–1675.
8. Sinclair DA, Guarente L. Extrachromosomal rDNA circles: a cause of aging in yeast. *Cell.* 1997;91(7):1033–1042.
9. Kaeberlein M, McVey M, Guarente L. The SIR2/3/4 complex and SIR2 alone promote longevity in *Saccharomyces cerevisiae* by two different mechanisms. *Genes Dev.* 1999;13(19):2570–2580.
10. Fabrizio P, Gattazzo C, Battistella L, Wei M, Cheng C, McGrew K, Longo VD. Sir2 blocks extreme life-span extension. *Cell.* 2005;123(4):655–667.
11. Houtkooper RH, Pirinen E, Auwerx J. Sirtuins as regulators of metabolism and healthspan. *Nat Rev Mol Cell Biol.* 2012;13(4):225–238.
12. Chang HC, Guarente L. SIRT1 and other sirtuins in metabolism. *Trends Endocrinol Metab.* 2014;25(3):138–145.
13. Masri S, Sassone-Corsi P. Sirtuins and the circadian clock: bridging chromatin and metabolism. *Sci Signal.* 2014;7(342):re6.

14. Tang BL. Sirt1's systemic protective roles and its promise as a target in antiaging medicine. *Transl Res.* 2011;157(5):276–284.
15. Hao CM, Haase VH. Sirtuins and their relevance to the kidney. *J Am Soc Nephrol.* 2010;21(10):1620–1627.
16. Someya S, Yu W, Hallows WC, Xu J, Vann JM, Leeuwenburgh C, Tanokura M, Denu JM, Prolla TA. Sirt3 mediates reduction of oxidative damage and prevention of age-related hearing loss under caloric restriction. *Cell.* 2010;143(5):802–812.
17. Mimura T, Kaji Y, Noma H, Funatsu H, Okamoto S. The role of SIRT1 in ocular aging. *Exp Eye Res.* 2013;116:17–26.
18. Lin SJ, Defossez PA, Guarente L. Requirement of NAD and SIR2 for life-span extension by calorie restriction in *Saccharomyces cerevisiae*. *Science.* 2000;289(5487):2126–2128.
19. Anderson RM, Bitterman KJ, Wood JG, Medvedik O, Sinclair DA. Nicotinamide and PNC1 govern lifespan extension by calorie restriction in *Saccharomycescerevisiae*. *Nature.* 2003;423(6936):181–185.
20. Cohen HY, Miller C, Bitterman KJ, Wall NR, Hekking B, Kessler B, Howitz KT, Gorospe M, de Cabo R, Sinclair DA. Calorie restriction promotes mammalian cell survival by inducing the SIRT1 deacetylase. *Science.* 2004;305(5682):390–392.
21. Chung KW, Kim DH, Park MH, Choi YJ, Kim ND, Lee J, Yu BP, Chung HY. Recent advances in calorie restriction research on aging. *Exp Gerontol.* 2013;48(10):1049–1053.
22. Gräff J, Kahn M, Samiei A, Gao J, Ota KT, Rei D, Tsai LH. A dietary regimen of caloric restriction or pharmacological activation of SIRT1 to delay the onset of neurodegeneration. *J Neurosci.* 2013;33(21):8951–8960.
23. Cantó C, Auwerx J. PGC-1alpha, SIRT1 and AMPK, an energy sensing network that controls energy expenditure. *Curr Opin Lipidol.* 2009;20(2):98–105.
24. Park S, Mori R, Shimokawa I. Do sirtuins promote mammalian longevity?. A critical review on its relevance to the longevity effect induced by calorie restriction. *Mol Cells.* 2013;35(6):474–480.
25. Guarente L. Calorie restriction and sirtuins revisited. *Genes Dev.* 2013;27(19):2072–2085.
26. Baur JA, Ungvari Z, Minor RK, Le Couteur DG, de Cabo R. Are sirtuins viable targets for improving healthspan and lifespan? *Nat Rev Drug Discov.* 2012;11(6):443–461.
27. Chakraborty C, Doss CGP. Sirtuins family: recent development as a drug target for aging, metabolism, and age related diseases. *Curr Drug Targets.* 2013;14(6):666–675.
28. Ingram DK, Roth GS. Calorie restriction mimetics: can you have your cake and eat it, too? *Ageing Res Rev.* 2015;20:46–62.
29. Tissenbaum HA, Guarente L. Increased dosage of a sir-2 gene extends lifespan in *Caenorhabditis elegans*. *Nature.* 2001;410(6825):227–230.
30. Rogina B, Helfand SL. Sir2 mediates longevity in the fly through a pathway related to calorie restriction. *Proc Natl Acad Sci USA.* 2004;101(45):15998–16003.
31. Wood JG, Rogina B, Lavu S, Howitz K, Helfand SL, Tatar M, Sinclair D. Sirtuin activators mimic caloric restriction and delay ageing in metazoans. *Nature.* 2004;430(7000):686–689.
32. Burnett C, Valentini S, Cabreiro F, Goss M, Somogyvári M, Piper MD, Hoddinott M, Sutphin GL, Leko V, McElwee JJ, Vazquez-Manrique RP, Orfila AM, Ackerman D, Au C, Vinti G, Riesen M, Howard K, Neri C, Bedalov A, Kaeberlein M, Soti C, Partridge L, Gems D. Absence of effects of Sir2 overexpression on lifespan in *C. elegans* and *Drosophila*. *Nature.* 2011;477(7365):482–485.
33. Viswanathan M, Guarente L. Regulation of *Caenorhabditis elegans* lifespan by sir-2.1 transgenes. *Nature.* 2011;477(7365):E1–E2.
34. Lombard DB, Pletcher SD, Cantó C, Auwerx J. Ageing: longevity hits a roadblock. *Nature.* 2011;477(7365):410–411.

35. Naiman S, Cohen HY. The contentious history of sirtuin debates. *Rambam Maimonides MedJ.* 2012;3(4). e0022.
36. Baur JA, Pearson KJ, Price NL, Jamieson HA, Lerin C, Kalra A, Prabhu VV, Allard JS, Lopez-Lluch G, Lewis K, Pistell PJ, Poosala S, Becker KG, Boss O, Gwinn D, Wang M, Ramaswamy S, Fishbein KW, Spencer RG, Lakatta EG, Le Couteur D, Shaw RJ, Navas P, Puigserver P, Ingram DK, de Cabo R, Sinclair DA. Resveratrol improves health and survival of mice on a high-calorie diet. *Nature.* 2006;444(7117):337–342.
37. Milne JC, Lambert PD, Schenk S, Carney DP, Smith JJ, Gagne DJ, Jin L, Boss O, Perni RB, Vu CB, Bemis JE, Xie R, Disch JS, Ng PY, Nunes JJ, Lynch AV, Yang H, Galonek H, Israelian K, Choy W, Iffland A, Lavu S, Medvedik O, Sinclair DA, Olefsky JM, Jirousek MR, Elliott PJ, Westphal CH. Small molecule activators of SIRT1 as therapeutics for the treatment of type 2 diabetes. *Nature.* 2007;450(7170):712–716.
38. Alcendor RR, Gao S, Zhai P, Zablocki D, Holle E, Yu X, Tian B, Wagner T, Vatner SF, Sadoshima J. Sirt1 regulates aging and resistance to oxidative stress in the heart. *Circ Res.* 2007;100(10):1512–1521.
39. Herranz D, Muñoz-Martin M, Cañamero M, Mulero F, Martinez-Pastor B, Fernandez-Capetillo O, Serrano M. Sirt1 improves healthy ageing and protects from metabolic syndrome-associated cancer. *Nat Commun.* 2010;1:3.
40. Kanfi Y, Naiman S, Amir G, Peshti V, Zinman G, Nahum L, Bar-Joseph Z, Cohen HY. The sirtuin SIRT6 regulates lifespan in male mice. *Nature.* 2012;483(7388):218–221.
41. North BJ, Rosenberg MA, Jeganathan KB, Hafner AV, Michan S, Dai J, Baker DJ, Cen Y, Wu LE, Sauve AA, van Deursen JM, Rosenzweig A, Sinclair DA. SIRT2 induces the checkpoint kinase BubR1 to increase lifespan. *EMBO J.* 2014;33(13):1438–1453.
42. Satoh A, Brace CS, Ben-Josef G, West T, Wozniak DF, Holtzman DM, Herzog ED, Imai SI. SIRT1 promotes the central adaptive response to diet restriction through activation of the dorsomedial and lateral nuclei of the hypothalamus. *J Neurosci.* 2010;30 (30):10220–10232.
43. Satoh A, Brace CS, Rensing N, Cliften P, Wozniak DF, Herzog ED, Yamada KA, Imai SI. Sirt1 extends life span and delays aging in mice through the regulation of Nk2 homeobox 1 in the DMH and LH. *Cell Metab.* 2013;18(3):416–430.
44. Mitchell SJ, Martin-Montalvo A, Mercken EM, Palacios HH, Ward TM, Abulwerdi G, Minor RK, Vlasuk GP, Ellis JL, Sinclair DA, Dawson J, Allison DB, Zhang Y, Becker KG, Bernier M, de Cabo R. The SIRT1 activator SRT1720 extends lifespan and improves health of mice fed a standard diet. *Cell Rep.* 2014;6(5):836–843.
45. Bishop NA, Lu T, Yankner BA. Neural mechanisms of ageing and cognitive decline. *Nature.* 2010;464(7288):529–535.
46. Brunet A, Sweeney LB, Sturgill JF, Chua KF, Greer PL, Lin Y, Tran H, Ross SE, Mostoslavsky R, Cohen HY, Hu LS, Cheng HL, Jedrychowski MP, Gygi SP, Sinclair DA, Alt FW, Greenberg ME. Stress-dependent regulation of FOXO transcription factors by the SIRT1 deacetylase. *Science.* 2004;303(5666):2011–2015.
47. Huang H, Tindall DJ. Dynamic FoxO transcription factors. *J Cell Sci.* 2007;120(Pt 15):2479–2487.
48. Luo J, Nikolaev AY, Imai S, Chen D, Su F, Shiloh A, Guarente L, Gu W. Negative control of p53 by Sir2alpha promotes cell survival under stress. *Cell.* 2001;107(2):137–148.
49. Hasegawa K, Yoshikawa K. Necdin regulates p53 acetylation via Sirtuin1 to modulate DNA damage response in cortical neurons. *J Neurosci.* 2008;28(35):8772–8784.
50. Yeung F, Hoberg JE, Ramsey CS, Keller MD, Jones DR, Frye RA, Mayo MW. Modulation of NF-kappaB-dependent transcription and cell survival by the SIRT1 deacetylase. *EMBO J.* 2004;23(12):2369–2380.
51. Chen J, Zhou Y, Mueller-Steiner S, Chen LF, Kwon H, Yi S, Mucke L, Gan L. SIRT1 protects against microglia-dependent amyloid-beta toxicity through inhibiting NF-kappaB signaling. *J Biol Chem.* 2005;280(48):40364–40374.

52. Kimura KD, Tissenbaum HA, Liu Y, Ruvkun G. daf-2, an insulin receptor-like gene that regulates longevity and diapause in *Caenorhabditis elegans*. *Science*. 1997;277 (5328):942–946.
53. Wolkow CA, Kimura KD, Lee MS, Ruvkun G. Regulation of *C. elegans* life-span by insulinlike signaling in the nervous system. *Science*. 2000;290(5489):147–150.
54. Cantó C, Gerhart-Hines Z, Feige JN, Lagouge M, Noriega L, Milne JC, Elliott PJ, Puigserver P, Auwerx J. AMPK regulates energy expenditure by modulating NAD+ metabolism and SIRT1 activity. *Nature*. 2009;458(7241):1056–1060.
55. Sugden MC, Caton PW, Holness MJ. PPAR control: it's SIRTainly as easy as PGC. *J Endocrinol*. 2010;204(2):93–104.
56. Dietrich MO, Antunes C, Geliang G, Liu ZW, Borok E, Nie Y, Xu AW, Souza DO, Gao Q, Diano S, Gao XB, Horvath TL. Agrp neurons mediate Sirt1's action on the melanocortin system and energy balance: roles for Sirt1 in neuronal firing and synaptic plasticity. *J Neurosci*. 2010;30(35):11815–11825.
57. Cyr NE, Steger JS, Toorie AM, Yang JZ, Stuart R, Nillni EA. Central Sirt1 regulates body weight and energy expenditure along with the POMC-derived peptide α-MSH and the processing enzyme CPE production in diet-induced obese male rats. *Endocrinology*. 2014;155(7):2423–2435.
58. Toorie AM, Nillni EA. Central Sirt1 regulates energy balance via the melanocortin system and alternate pathways. *Mol Endocrinol*. 2014;28(9):1423–1434.
59. Ng F, Wijaya L, Tang BL. SIRT1 in the brain-connections with aging-associated disorders and lifespan. *Front Cell Neurosci*. 2015;9:64.
60. Maury E, Ramsey KM, Bass J. Circadian rhythms and metabolic syndrome: from experimental genetics to human disease. *Circ Res*. 2010;106(3):447–462.
61. Rey G, Reddy AB. Connecting cellular metabolism to circadian clocks. *Trends Cell Biol*. 2013;23(5):234–241.
62. Nakahata Y, Kaluzova M, Grimaldi B, Sahar S, Hirayama J, Chen D, Guarente LP, Sassone-Corsi P. The NAD+-dependent deacetylase SIRT1 modulates CLOCK-mediated chromatin remodeling and circadian control. *Cell*. 2008;134(2): 329–340.
63. Michán S, Li Y, Chou MMH, Parrella E, Ge H, Long JM, Allard JS, Lewis K, Miller M, Xu W, Mervis RF, Chen J, Guerin KI, Smith LEH, McBurney MW, Sinclair DA, Baudry M, de Cabo R, Longo VD. SIRT1 is essential for normal cognitive function and synaptic plasticity. *J Neurosci*. 2010;30(29):9695–9707.
64. Gao J, Wang WY, Mao YW, Gräff J, Guan JS, Pan L, Mak G, Kim D, Su SC, Tsai LH. A novel pathway regulates memory and plasticity via SIRT1 and miR-134. *Nature*. 2010;466(7310):1105–1109.
65. Prozorovski T, Schulze-Topphoff U, Glumm R, Baumgart J, Schröter F, Ninnemann O, Siegert E, Bendix I, Brüstle O, Nitsch R, Zipp F, Aktas O. Sirt1 contributes critically to the redox-dependent fate of neural progenitors. *Nat Cell Biol*. 2008;10 (4):385–394.
66. Tang BL. Sirt1's complex roles in neuroprotection. *Cell Mol Neurobiol*. 2009;29 (8):1093–1103.
67. Albani D, Polito L, Forloni G. Sirtuins as novel targets for Alzheimer's disease and other neurodegenerative disorders: experimental and genetic evidence. *J Alzheimers Dis*. 2010;19(1):11–26.
68. Kincaid B, Bossy-Wetzel E. Forever young: SIRT3 a shield against mitochondrial meltdown, aging, and neurodegeneration. *Front Aging Neurosci*. 2013;5:48.
69. Outeiro TF, Kontopoulos E, Altmann SM, Kufareva I, Strathearn KE, Amore AM, Volk CB, Maxwell MM, Rochet JC, McLean PJ, Young AB, Abagyan R, Feany MB, Hyman BT, Kazantsev AG. Sirtuin 2 inhibitors rescue alpha-synuclein-mediated toxicity in models of Parkinson's disease. *Science*. 2007;317(5837):516–519.

70. Luo J, Su F, Chen D, Shiloh A, Gu W. Deacetylation of p53 modulates its effect on cell growth and apoptosis. *Nature*. 2000;408(6810):377–381.
71. Hasegawa K, Wakino S, Yoshioka K, Tatematsu S, Hara Y, Minakuchi H, Washida N, Tokuyama H, Hayashi K, Itoh H. Sirt1 protects against oxidative stress-induced renal tubular cell apoptosis by the bidirectional regulation of catalase expression. *Biochem Biophys Res Commun*. 2008;372(1):51–56.
72. Hernández-Jiménez M, Hurtado O, Cuartero MI, Ballesteros I, Moraga A, Pradillo JM, McBurney MW, Lizasoain I, Moro MA. Silent information regulator 1 protects the brain against cerebral ischemic damage. *Stroke*. 2013;44(8):2333–2337.
73. Xiong H, Pang J, Yang H, Dai M, Liu Y, Ou Y, Huang Q, Chen S, Zhang Z, Xu Y, Lai L, Zheng Y. Activation of miR-34a/SIRT1/p53 signaling contributes to cochlear hair cell apoptosis: implications for age-related hearing loss. *Neurobiol Aging*. 2015;36 (4):1692–1701.
74. Zuccato C, Cattaneo E. Brain-derived neurotrophic factor in neurodegenerative diseases. *Nat Rev Neurol*. 2009;5(6):311–322.
75. Raval AP, Dave KR, Pérez-Pinzón MA. Resveratrol mimics ischemic preconditioning in the brain. *J Cereb Blood Flow Metab*. 2006;26(9):1141–1147.
76. Koronowski KB, Dave KR, Saul I, Camarena V, Thompson JW, Neumann JT, Young JI, Perez-Pinzon MA. Resveratrol preconditioning induces a novel extended window of ischemic tolerance in the mouse brain. *Stroke*. 2015;46(8):2293–2298.
77. Yang Y, Jiang S, Dong Y, Fan C, Zhao L, Yang X, Li J, Di S, Yue L, Liang G, Reiter RJ, Qu Y. Melatonin prevents cell death and mitochondrial dysfunction via a SIRT1-dependent mechanism during ischemic-stroke in mice. *J Pineal Res*. 2015;58(1):61–70.
78. Zhao Y, Luo P, Guo Q, Li S, Zhang L, Zhao M, Xu H, Yang Y, Poon W, Fei Z. Interactions between SIRT1 and MAPK/ERK regulate neuronal apoptosis induced by traumatic brain injury in vitro and in vivo. *Exp Neurol*. 2012;237(2):489–498.
79. Ng F, Tang BL. When is Sirt1 activity bad for dying neurons? *Front Cell Neurosci*. 2013;7:186.
80. Liu D, Gharavi R, Pitta M, Gleichmann M, Mattson MP. Nicotinamide prevents NAD + depletion and protects neurons against excitotoxicity and cerebral ischemia: NAD+ consumption by SIRT1 may endanger energetically compromised neurons. *Neuromolecular Med*. 2009;11(1):28–42.
81. Sansone L, Reali V, Pellegrini L, Villanova L, Aventaggiato M, Marfe G, Rosa R, Nebbioso M, Tafani M, Fini M, Russo MA, Pucci B. SIRT1 silencing confers neuroprotection through IGF-1 pathway activation. *J Cell Physiol*. 2013;228(8):1754–1761.
82. Yin J, Han P, Tang Z, Liu Q, Shi J. Sirtuin 3 mediates neuroprotection of ketones against ischemic stroke. *J Cereb Blood Flow Metab*. 2015;35(11):1783–1789.
83. Dai SH, Chen T, Li X, Yue KY, Luo P, Yang LK, Zhu J, Wang YH, Fei Z, Jiang XF. Sirt3 confers protection against neuronal ischemia by inducing autophagy: involvement of the AMPK-mTOR pathway. *Free Radic Biol Med*. 2017;108:345–353.
84. Krey L, Lühder F, Kusch K, Czech-Zechmeister B, Könnecke B, Fleming Outeiro T, Trendelenburg G. Knockout of silent information regulator 2 (SIRT2) preserves neurological function after experimental stroke in mice. *J Cereb Blood Flow Metab*. 2015;35 (12):2080–2088.
85. Xie XQ, Zhang P, Tian B, Chen XQ. Downregulation of NAD-dependent deacetylase SIRT2 protects mouse brain against ischemic stroke. *Mol Neurobiol*. 2016.
86. Alzheimer's Association.Alzheimer's disease facts and figures. *Alzheimers Dement*. 2015;11(3):332–384.
87. Fiest KM, Roberts JI, Maxwell CJ, Hogan DB, Smith EE, Frolkis A, Cohen A, Kirk A, Pearson D, Pringsheim T, Venegas-Torres A, Jetté N. The prevalence and incidence of dementia due to Alzheimer's disease: a systematic review and meta-analysis. *Can J Neurol Sci*. 2016;43(Suppl 1):S51–S82.

88. Scheltens P, Blennow K, Breteler MMB, de Strooper B, Frisoni GB, Salloway S, Van der Flier WM. Alzheimer's disease. *Lancet.* 2016;388(10043):505–517.
89. Ittner LM, Götz J. Amyloid-β and tau: a toxic pas de deux in Alzheimer's disease. *Nat Rev Neurosci.* 2011;12(2):65–72.
90. Bloom GS. Amyloid-β and tau: the trigger and bullet in Alzheimer disease pathogenesis. *JAMA Neurol.* 2014;71(4):505–508.
91. Vassar R, Kovacs DM, Yan R, Wong PC. The beta-secretase enzyme BACE in health and Alzheimer's disease: regulation, cell biology, function, and therapeutic potential. *J Neurosci.* 2009;29(41):12787–12794.
92. De Strooper B, Iwatsubo T, Wolfe MS. Presenilins and γ-secretase: structure, function, and role in Alzheimer disease. *Cold Spring Harb Perspect Med.* 2012;2(1). a006304.
93. Karran E, Mercken M, De Strooper B. The amyloid cascade hypothesis for Alzheimer's disease: an appraisal for the development of therapeutics. *Nat Rev Drug Discov.* 2011;10 (9):698–712.
94. Nisbet RM, Polanco JC, Ittner LM, Götz J. Tau aggregation and its interplay with amyloid-β. *Acta Neuropathol.* 2015;129(2):207–220.
95. Tanzi RE. The genetics of Alzheimer disease. *Cold Spring Harb Perspect Med.* 2012;2(10): a006296.
96. Julien C, Tremblay C, Emond V, Lebbadi M, Salem N, Bennett DA, Calon F. Sirtuin 1 reduction parallels the accumulation of tau in Alzheimer disease. *J Neuropathol Exp Neurol.* 2009;68(1):48–58.
97. Lutz MI, Milenkovic I, Regelsberger G, Kovacs GG. Distinct patterns of sirtuin expression during progression of Alzheimer's disease. *Neuromolecular Med.* 2014;16 (2):405–414.
98. Feng X, Liang N, Zhu D, Gao Q, Peng L, Dong H, Yue Q, Liu H, Bao L, Zhang J, Hao J, Gao Y, Yu X, Sun J. Resveratrol inhibits β-amyloid-induced neuronal apoptosis through regulation of SIRT1-ROCK1 signaling pathway. *PLoS One.* 2013;8(3): e59888.
99. Lee HR, Shin HK, Park SY, Kim HY, Lee WS, Rhim BY, Hong KW, Kim CD. Cilostazol suppresses β-amyloid production by activating a disintegrin and metalloproteinase 10 via the upregulation of SIRT1-coupled retinoic acid receptor-β. *J Neurosci Res.* 2014;92(11):1581–1590.
100. Marwarha G, Raza S, Meiers C, Ghribi O. Leptin attenuates BACE1 expression and amyloid-(genesis via the activation of SIRT1 signaling pathway. *Biochim Biophys Acta.* 2014;1842(9):1587–1595.
101. Sun Q, Jia N, Wang W, Jin H, Xu J, Hu H. Activation of SIRT1 by curcumin blocks the neurotoxicity of amyloid-β25-35 in rat cortical neurons. *Biochem Biophys Res Commun.* 2014;448(1):89–94.
102. Kim D, Nguyen MD, Dobbin MM, Fischer A, Sananbenesi F, Rodgers JT, Delalle I, Baur JA, Sui G, Armour SM, Puigserver P, Sinclair DA, Tsai LH. SIRT1 deacetylase protects against neurodegeneration in models for Alzheimer's disease and amyotrophic lateral sclerosis. *EMBO J.* 2007;26(13):3169–3179.
103. Porquet D, Casadesús G, Bayod S, Vicente A, Canudas AM, Vilaplana J, Pelegrí C, Sanfeliu C, Camins A, Pallàs M, del Valle J. Dietary resveratrol prevents Alzheimer's markers and increases life span in SAMP8. *Age.* 2013;35(5):1851–1865.
104. Du LL, Xie JZ, Cheng XS, Li XH, Kong FL, Jiang X, Ma ZW, Wang JZ, Chen C, Zhou XW. Activation of sirtuin 1 attenuates cerebral ventricular streptozotocin-induced tau hyperphosphorylation and cognitive injuries in rat hippocampi. *Age.* 2014;36 (2):613–623.
105. Porquet D, Griñán-Ferré C, Ferrer I, Camins A, Sanfeliu C, Del Valle J, Pallàs M. Neuroprotective role of trans-resveratrol in a murine model of familial Alzheimer's disease. *J Alzheimers Dis.* 2014;42(4):1209–1220.

106. Gao J, Zhou R, You X, Luo F, He H, Chang X, Zhu L, Ding X, Yan T. Salidroside suppresses inflammation in a D-galactose-induced rat model of Alzheimer's disease via SIRT1/NF-κB pathway. *Metab Brain Dis.* 2016;31(4):771–778.
107. Shah SA, Yoon GH, Chung SS, Abid MN, Kim TH, Lee HY, Kim MO. Novel osmotin inhibits SREBP2 via the AdipoR1/AMPK/SIRT1 pathway to improve Alzheimer's disease neuropathological deficits. *Mol Psychiatry.* 2016;22(3):407–416.
108. Jung ES, Choi H, Song H, Hwang YJ, Kim A, Ryu H, Mook-Jung I. p53-dependent SIRT6 expression protects Aβ42-induced DNA damage. *Sci Rep.* 2016;6:25628.
109. Wong SY, Tang BL. SIRT1 as a therapeutic target for Alzheimer's disease. *Rev Neurosci.* 2016;27(8):813–825.
110. Tang BL. Alzheimer's disease: channeling APP to non-amyloidogenic processing. *Biochem Biophys Res Commun.* 2005;331(2):375–378.
111. Endres K, Fahrenholz F. The Role of the anti-amyloidogenic secretase ADAM10 in shedding the APP-like proteins. *Curr Alzheimer Res.* 2012;9(2):157–164.
112. Qin W, Yang T, Ho L, Zhao Z, Wang J, Chen L, Zhao W, Thiyagarajan M, MacGrogan D, Rodgers JT, Puigserver P, Sadoshima J, Deng H, Pedrini S, Gandy S, Sauve AA, Pasinetti GM. Neuronal SIRT1 activation as a novel mechanism underlying the prevention of Alzheimer disease amyloid neuropathology by calorie restriction. *J Biol Chem.* 2006;281(31):21745–21754.
113. Theendakara V, Patent A, Peters Libeu CA, Philpot B, Flores S, Descamps O, Poksay KS, Zhang Q, Cailing G, Hart M, John V, Rao RV, Bredesen DE. Neuroprotective Sirtuin ratio reversed by ApoE4. *Proc Natl Acad Sci USA.* 2013;110(45):18303–18308.
114. Irwin DJ, Cohen TJ, Grossman M, Arnold SE, McCarty-Wood E, Van Deerlin VM, Lee VMY, Trojanowski JQ. Acetylated tau neuropathology in sporadic and hereditary tauopathies. *Am J Pathol.* 2013;183(2):344–351.
115. Min SW, Chen X, Tracy TE, Li Y, Zhou Y, Wang C, Shirakawa K, Minami SS, Defensor E, Mok SA, Sohn PD, Schilling B, Cong X, Ellerby L, Gibson BW, Johnson J, Krogan N, Shamloo M, Gestwicki J, Masliah E, Verdin E, Gan L. Critical role of acetylation in tau-mediated neurodegeneration and cognitive deficits. *Nat Med.* 2015;21(10):1154–1162.
116. Min SW, Cho SH, Zhou Y, Schroeder S, Haroutunian V, Seeley WW, Huang EJ, Shen Y, Masliah E, Mukherjee C, Meyers D, Cole PA, Ott M, Gan L. Acetylation of tau inhibits its degradation and contributes to tauopathy. *Neuron.* 2010;67(6):953–966.
117. Ng F, Tang BL. Sirtuins' modulation of autophagy. *J Cell Physiol.* 2013;228(12):2262–2270.
118. Chang C, Su H, Zhang D, Wang Y, Shen Q, Liu B, Huang R, Zhou T, Peng C, Wong CCL, Shen HM, Lippincott-Schwartz J, Liu W. AMPK-dependent phosphorylation of GAPDH triggers Sirt1 activation and is necessary for autophagy upon glucose starvation. *Mol Cell.* 2015;60(6):930–940.
119. Miners JS, Baig S, Palmer J, Palmer LE, Kehoe PG, Love S. Abeta-degrading enzymes in Alzheimer's disease. *Brain Pathol.* 2008;18(2):240–252.
120. Qiu WQ, Folstein MF. Insulin, insulin-degrading enzyme and amyloid-beta peptide in Alzheimer's disease: review and hypothesis. *Neurobiol Aging.* 2006;27(2):190–198.
121. Haque R, Nazir A. Insulin-degrading enzyme: a link between Alzheimer's and type 2 diabetes mellitus. *CNS Neurol Disord Drug Targets.* 2014;13(2):259–264.
122. El-Sayed NS, Bayan Y. Possible role of resveratrol targeting estradiol and neprilysin pathways in lipopolysaccharide model of Alzheimer disease. *Adv Exp Med Biol.* 2015;822:107–118.
123. Kalia LV, Lang AE. Parkinson's disease. *Lancet.* 2015;386(9996):896–912.
124. Halliday G, Lees A, Stern M. Milestones in Parkinson's disease: clinical and pathologic features. *Mov Disord.* 2011;26(6):1015–1021.

125. Bellucci A, Mercuri NB, Venneri A, Faustini G, Longhena F, Pizzi M, Missale C, Spano P. Parkinson's disease: from synaptic loss to connectome dysfunction. *Neuropathol Appl Neurobiol.* 2016;42(1):77–94.
126. Klingelhoefer L, Reichmann H. Pathogenesis of Parkinson disease: the gut-brain axis and environmental factors. *Nat Rev Neurol.* 2015;11(11):625–636.
127. Shulman JM, De Jager PL, Feany MB. Parkinson's disease: genetics and pathogenesis. *Annu Rev Pathol.* 2011;6:193–222.
128. Klein C, Westenberger A. Genetics of Parkinson's disease. *Cold Spring Harb Perspect Med.* 2012;2(1):a008888.
129. Benskey MJ, Perez RG. Manfredsson FP. The contribution of alpha synuclein to neuronal survival and function—implications for Parkinson's disease. *J Neurochem.* 2016;137(3):331–359.
130. Caulfield TR, Fiesel FC, Springer W. Activation of the E3 ubiquitin ligase Parkin. *Biochem Soc Trans.* 2015;43(2):269–274.
131. Aerts L, De Strooper B, Morais VA. PINK1 activation-turning on a promiscuous kinase. *Biochem Soc Trans.* 2015;43(2):280–286.
132. Ariga H, Takahashi-Niki K, Kato I, Maita H, Niki T, Iguchi-Ariga SMM. Neuroprotective function of DJ-1 in Parkinson's disease. *Oxid Med Cell Longev.* 2013;2013:683920.
133. Kalia LV, Kalia SK. α-Synuclein and Lewy pathology in Parkinson's disease. *Curr Opin Neurol.* 2015;28(4):375–381.
134. Atik A, Stewart T, Zhang J. Alpha-synuclein as a biomarker for Parkinson's disease. *Brain Pathol.* 2016;26(3):410–418.
135. Tang BL. Sirtuins as modifiers of Parkinson's disease pathology. *J Neurosci Res.* 2017;95 (4):930–942.
136. Albani D, Polito L, Batelli S, De Mauro S, Fracasso C, Martelli G, Colombo L, Manzoni C, Salmona M, Caccia S, Negro A, Forloni G. The SIRT1 activator resveratrol protects SK-N-BE cells from oxidative stress and against toxicity caused by alpha-synuclein or amyloid-beta (1-42) peptide. *J Neurochem.* 2009;110(5):1445–1456.
137. Wu Y, Li X, Zhu JX, Xie W, Le W, Fan Z, Jankovic J, Pan T. Resveratrol-activated AMPK/SIRT1/autophagy in cellular models of Parkinson's disease. *Neurosignals.* 2011;19(3):163–174.
138. Mudò G, Mäkelä J, Di Liberto V, Tselykh TV, Olivieri M, Piepponen P, Eriksson O, Mälkiä A, Bonomo A, Kairisalo M, Aguirre JA, Korhonen L, Belluardo N, Lindholm D. Transgenic expression and activation of PGC-1α protect dopaminergic neurons in the MPTP mouse model of Parkinson's disease. *Cell Mol Life Sci.* 2012;69(7):1153–1165.
139. Chao J, Yu MS, Ho YS, Wang M, Chang RCC. Dietary oxyresveratrol prevents parkinsonian mimetic 6-hydroxydopamine neurotoxicity. *Free Radic Biol Med.* 2008;45 (7):1019–1026.
140. Chao J, Li H, Cheng KW, Yu MS, Chang RCC, Wang M. Protective effects of pinostilbene, a resveratrol methylated derivative, against 6-hydroxydopamine-induced neurotoxicity in SH-SY5Y cells. *J Nutr Biochem.* 2010;21(6):482–489.
141. Zou XD, Guo SQ, Hu ZW, Li WL. NAMPT protects against 6-hydroxydopamine-induced neurotoxicity in PC12 cells through modulating SIRT1 activity. *Mol Med Rep.* 2016;13(5):4058–4064.
142. van Ham TJ, Thijssen KL, Breitling R, Hofstra RMW, Plasterk RHA, Nollen EAA. *C. elegans* model identifies genetic modifiers of alpha-synuclein inclusion formation during aging. *PLoS Genet.* 2008;4(3):e1000027.
143. Hu W, Guan LS, Dang XB, Ren PY, Zhang YL. Small-molecule inhibitors at the PSD-95/nNOS interface attenuate MPP+-induced neuronal injury through Sirt3 mediated inhibition of mitochondrial dysfunction. *Neurochem Int.* 2014;79:57–64.
144. Zhang JY, Deng YN, Zhang M, Su H, Qu QM. SIRT3 acts as a neuroprotective agent in rotenone-induced Parkinson cell model. *Neurochem Res.* 2016;41(7):1761–1763.

145. Liu L, Peritore C, Ginsberg J, Kayhan M, Donmez G. SIRT3 attenuates MPTP-induced nigrostriatal degeneration via enhancing mitochondrial antioxidant capacity. *Neurochem Res.* 2015;40(3):600–608.
146. Liu L, Peritore C, Ginsberg J, Shih J, Arun S, Donmez G. Protective role of SIRT5 against motor deficit and dopaminergic degeneration in MPTP-induced mice model of Parkinson's disease. *Behav Brain Res.* 2015;281:215–221.
147. Di Fruscia P, Zacharioudakis E, Liu C, Moniot S, Laohasinnarong S, Khongkow M, Harrison IF, Koltsida K, Reynolds CR, Schmidtkunz K, Jung M, Chapman KL, Steegborn C, Dexter DT, Sternberg MJE, Lam EWF, Fuchter MJ. The discovery of a highly selective 5,6,7,8-tetrahydrobenzo[4,5]thieno[2,3-d]pyrimidin-4(3H)-one SIRT2 inhibitor that is neuroprotective in an in vitro Parkinson's disease model. *ChemMedChem.* 2015;10(1):69–82.
148. Chen X, Wales P, Quinti L, Zuo F, Moniot S, Herisson F, Rauf NA, Wang H, Silverman RB, Ayata C, Maxwell MM, Steegborn C, Schwarzschild MA, Outeiro TF, Kazantsev AG. The sirtuin-2 inhibitor AK7 is neuroprotective in models of Parkinson's disease but not amyotrophic lateral sclerosis and cerebral ischemia. *PLoS One.* 2015;10(1): e0116919.
149. Liu L, Arun A, Ellis L, Peritore C, Donmez G. SIRT2 enhances 1-methyl-4-phenyl-1,2,3,6-tetrahydropyridine (MPTP)-induced nigrostriatal damage via apoptotic pathway. *Front Aging Neurosci.* 2014;6:184.
150. Büeler H. Impaired mitochondrial dynamics and function in the pathogenesis of Parkinson's disease. *Exp Neurol.* 2009;218(2):235–246.
151. Trancikova A, Tsika E, Moore DJ. Mitochondrial dysfunction in genetic animal models of Parkinson's disease. *Antioxid Redox Signal.* 2012;16(9):896–919.
152. Van Laar VS, Berman SB. The interplay of neuronal mitochondrial dynamics and bioenergetics: implications for Parkinson's disease. *Neurobiol Dis.* 2013;51:43–55.
153. Khan RS, Fonseca-Kelly Z, Callinan C, Zuo L, Sachdeva MM, Shindler KS. SIRT1 activating compounds reduce oxidative stress and prevent cell death in neuronal cells. *Front Cell Neurosci.* 2012;6:63.
154. Baldelli S, Aquilano K, Ciriolo MR. PGC-1α buffers ROS-mediated removal of mitochondria during myogenesis. *Cell Death Dis.* 2014;5(11):e1515.
155. Wang SJ, Zhao XH, Chen W, Bo N, Wang XJ, Chi ZF, Wu W. Sirtuin 1 activation enhances the PGC-1α/mitochondrial antioxidant system pathway in status epilepticus. *Mol Med Rep.* 2015;11(1):521–526.
156. Tan M, Tang C, Zhang Y, Cheng Y, Cai L, Chen X, Gao Y, Deng Y, Pan M. SIRT1/PGC-1α signaling protects hepatocytes against mitochondrial oxidative stress induced by bile acids. *Free Radic Res.* 2015;49(8):935–945.
157. Barker RA, Williams-Gray CH. The spectrum of clinical features seen with alpha synuclein pathology. *Neuropathol Appl Neurobiol.* 2016;42(1):6–19.
158. Westerheide SD, Anckar J, Stevens SM, Sistonen L, Morimoto RI. Stress-inducible regulation of heat shock factor 1 by the deacetylase SIRT1. *Science.* 2009;323 (5917):1063–1066.
159. Raynes R, Pombier KM, Nguyen K, Brunquell J, Mendez JE, Westerheide SD. The SIRT1 modulators AROS and DBC1 regulate HSF1 activity and the heat shock response. *PLoS One.* 2013;8(1):e54364.
160. Tomita T, Hamazaki J, Hirayama S, McBurney MW, Yashiroda H, Murata S. Sirt1-deficiency causes defective protein quality control. *Sci Rep.* 2015;5:12613.
161. Perdiz D, Mackeh R, Poüs C, Baillet A. The ins and outs of tubulin acetylation: more than just a post-translational modification? *Cell Signal.* 2011;23(5):763–771.
162. Cai ZL, Shi JJ, Yang YP, Cao BY, Wang F, Huang JZ, Yang F, Zhang P, Liu CF. MPP+ impairs autophagic clearance of alpha-synuclein by impairing the activity of dynein. *Neuroreport.* 2009;20(6):569–573.

163. Cappelletti G, Casagrande F, Calogero A, De Gregorio C, Pezzoli G, Cartelli D. Linking microtubules to Parkinson's disease: the case of parkin. *Biochem Soc Trans.* 2015;43 (2):292–296.
164. Arduíno DM, Esteves AR, Cortes L, Silva DF, Patel B, Grazina M, Swerdlow RH, Oliveira CR, Cardoso SM. Mitochondrial metabolism in Parkinson's disease impairs quality control autophagy by hampering microtubule-dependent traffic. *Hum Mol Genet.* 2012;21(21):4680–4702.
165. Esteves AR, Gozes I, Cardoso SM. The rescue of microtubule-dependent traffic recovers mitochondrial function in Parkinson's disease. *Biochim Biophys Acta.* 2014;1842(1):7–21.
166. Godena VK, Brookes-Hocking N, Moller A, Shaw G, Oswald M, Sancho RM, Miller CCJ, Whitworth AJ, De Vos KJ. Increasing microtubule acetylation rescues axonal transport and locomotor deficits caused by LRRK2 Roc-COR domain mutations. *Nat Commun.* 2014;5:5245.
167. Riva N, Agosta F, Lunetta C, Filippi M, Quattrini A. Recent advances in amyotrophic lateral sclerosis. *J Neurol.* 2016;263(6):1241–1254.
168. Renton AE, Chiò A, Traynor BJ. State of play in amyotrophic lateral sclerosis genetics. *Nat Neurosci.* 2014;17(1):17–23.
169. Jaarsma D, Haasdijk ED, Grashorn JA, Hawkins R, van Duijn W, Verspaget HW, London J, Holstege JC. Human Cu/Zn superoxide dismutase (SOD1) overexpression in mice causes mitochondrial vacuolization, axonal degeneration, and premature motoneuron death and accelerates motoneuron disease in mice expressing a familial amyotrophic lateral sclerosis mutant SOD1. *Neurobiol Dis.* 2000;7(6 Pt B):623–643.
170. Song W, Song Y, Kincaid B, Bossy B, Bossy-Wetzel E. Mutant SOD1G93A triggers mitochondrial fragmentation in spinal cord motor neurons: neuroprotection by SIRT3 and PGC-1. *Neurobiol Dis.* 2013;51:72–81.
171. Yamanaka K, Chun SJ, Boillee S, Fujimori-Tonou N, Yamashita H, Gutmann DH, Takahashi R, Misawa H, Cleveland DW. Astrocytes as determinants of disease progression in inherited amyotrophic lateral sclerosis. *Nat Neurosci.* 2008;11(3):251–253.
172. Harlan BA, Pehar M, Sharma DR, Beeson G, Beeson CC, Vargas MR. Enhancing NAD+ salvage pathway reverts the toxicity of primary astrocytes expressing amyotrophic lateral sclerosis-linked mutant superoxide dismutase 1 (SOD1). *J Biol Chem.* 2016;291(20):10836–10846.
173. Han S, Choi JR, Soon Shin K, Kang SJ. Resveratrol upregulated heat shock proteins and extended the survival of G93A-SOD1 mice. *Brain Res.* 2012;1483:112–117.
174. Mancuso R, del Valle J, Modol L, Martinez A, Granado-Serrano AB, Ramirez-Núñez O, Pallás M, Portero-Otin M, Osta R, Navarro X. Resveratrol improves motoneuron function and extends survival in SOD1(G93A) ALS mice. *Neurotherapeutics.* 2014;11 (2):419–432.
175. Watanabe S, Ageta-Ishihara N, Nagatsu S, Takao K, Komine O, Endo F, Miyakawa T, Misawa H, Takahashi R, Kinoshita M, Yamanaka K. SIRT1 overexpression ameliorates a mouse model of SOD1-linked amyotrophic lateral sclerosis via HSF1/HSP70i chaperone system. *Molecular Brain.* 2014;7:62.
176. Zhao W, Varghese M, Yemul S, Pan Y, Cheng A, Marano P, Hassan S, Vempati P, Chen F, Qian X, Pasinetti GM. Peroxisome proliferator activator receptor gamma coactivator-1alpha (PGC-1α) improves motor performance and survival in a mouse model of amyotrophic lateral sclerosis. *Mol Neurodegener.* 2011;6(1):51.
177. Tang BL. Could sirtuin activities modify ALS onset and progression? *Cell Mol Neurobiol.* 2017;37(7):1147–1160.
178. Seo JS, Moon MH, Jeong JK, Seol JW, Lee YJ, Park BH, Park SY. SIRT1, a histone deacetylase, regulates prion protein-induced neuronal cell death. *Neurobiol Aging.* 2012;33(6):1110–1120.

179. Jeong JK, Moon MH, Lee YJ, Seol JW, Park SY. Autophagy induced by the class III histone deacetylase Sirt1 prevents prion peptide neurotoxicity. *Neurobiol Aging.* 2013;34 (1):146–156.
180. Walker FO. Huntington's disease. *Lancet.* 2007;369(9557):218–228.
181. Bates GP, Dorsey R, Gusella JF, Hayden MR, Kay C, Leavitt BR, Nance M, Ross CA, Scahill RI, Wetzel R, Wild EJ, Tabrizi SJ. Huntington disease. *Nat Rev Dis Primers.* 2015;1:15005.
182. Ross CA, Tabrizi SJ. Huntington's disease: from molecular pathogenesis to clinical treatment. *Lancet Neurol.* 2011;10(1):83–98.
183. The Huntington's Disease Collaborative Research Group.A novel gene containing a trinucleotide repeat that is expanded and unstable on Huntington's disease chromosomes. *Cell.* 1993;72(6):971–983.
184. Saudou F, Humbert S. The biology of Huntingtin. *Neuron.* 2016;89(5):910–926.
185. Orr HT. Polyglutamine neurodegeneration: expanded glutamines enhance native functions. *Curr Opin Genet Dev.* 2012;22(3):251–255.
186. Huntington Study Group.Unified Huntington's disease rating scale: reliability and consistency. *Mov Disord.* 1996;11(2):136–142.
187. Aylward EH, Nopoulos PC, Ross CA, Langbehn DR, Pierson RK, Mills JA, Johnson HJ, Magnotta VA, Juhl AR, Paulsen JS. PREDICT-HD investigators and coordinators of Huntington study group. Longitudinal change in regional brain volumes in prodromal Huntington disease. *J Neurol Neurosurg Psychiatry.* 2011;82(4):405–410.
188. Mason SL, Barker RA. Advancing pharmacotherapy for treating Huntington's disease: a review of the existing literature. *Expert Opin Pharmacother.* 2016;17(1):41–52.
189. Tartari M, Gissi C, Lo Sardo V, Zuccato C, Picardi E, Pesole G, Cattaneo E. Phylogenetic comparison of Huntingtin homologues reveals the appearance of a primitive polyQ in sea urchin. *Mol Biol Evol.* 2008;25(2):330–338.
190. Andrade MA, Bork P. HEAT repeats in the Huntington's disease protein. *Nat Genet.* 1995;11(2):115–116.
191. Palidwor GA, Shcherbinin S, Huska MR, Rasko T, Stelzl U, Arumughan A, Foulle R, Porras P, Sanchez-Pulido L, Wanker EE, Andrade-Navarro MA. Detection of alpha-rod protein repeats using a neural network and application to Huntingtin. *PLoS Comput Biol.* 2009;5(3):e1000304.
192. Seong IS, Woda JM, Song JJ, Lloret A, Abeyrathne PD, Woo CJ, Gregory G, Lee JM, Wheeler VC, Walz T, Kingston RE, Gusella JF, Conlon RA, MacDonald ME. Huntingtin facilitates polycomb repressive complex 2. *Hum Mol Genet.* 2010;19 (4):573–583.
193. Ochaba J, Lukacsovich T, Csikos G, Zheng S, Margulis J, Salazar L, Mao K, Lau AL, Yeung SY, Humbert S, Saudou F, Klionsky DJ, Finkbeiner S, Zeitlin SO, Marsh JL, Housman DE, Thompson LM, Steffan JS. Potential function for the Huntingtin protein as a scaffold for selective autophagy. *Proc Natl Acad Sci USA.* 2014;111(47): 16889–16894.
194. Rui YN, Xu Z, Patel B, Chen Z, Chen D, Tito A, David G, Sun Y, Stimming EF, Bellen HJ, Cuervo AM, Zhang S. Huntingtin functions as a scaffold for selective macroautophagy. *Nat Cell Biol.* 2015;17(3):262–275.
195. Steffan JS, Kazantsev A, Spasic-Boskovic O, Greenwald M, Zhu YZ, Gohler H, Wanker EE, Bates GP, Housman DE, Thompson LM. The Huntington's disease protein interacts with p53 and CREB-binding protein and represses transcription. *Proc Natl Acad Sci USA.* 2000;97(12):6763–6768.
196. Li SH, Cheng AL, Zhou H, Lam S, Rao M, Li H, Li XJ. Interaction of Huntington disease protein with transcriptional activator Sp1. *Mol Cell Biol.* 2002;22(5):1277–1287.
197. Zuccato C, Tartari M, Crotti A, Goffredo D, Valenza M, Conti L, Cataudella T, Leavitt BR, Hayden MR, Timmusk T, Rigamonti D, Cattaneo E. Huntingtin interacts with

REST/NRSF to modulate the transcription of NRSE-controlled neuronal genes. *Nat Genet.* 2003;35(1):76–83.
198. Wu LLY, Zhou XF. Huntingtin associated protein 1 and its functions. *Cell Adh Migr.* 2009;3(1):71–76.
199. Mackenzie KD, Lim Y, Duffield MD, Chataway T, Zhou XF, Keating DJ. Huntingtin-associated protein-1 (HAP1) regulates endocytosis and interacts with multiple trafficking-related proteins. *Cell Signal.* 2017;35:176–187.
200. Bhattacharyya NP, Banerjee M, Majumder P. Huntington's disease: roles of huntingtin-interacting protein 1 (HIP-1) and its molecular partner HIPPI in the regulation of apoptosis and transcription. *FEBS J.* 2008;275(17):4271–4279.
201. McGuire JR, Rong J, Li SH, Li XJ. Interaction of Huntingtin-associated protein-1 with kinesin light chain: implications in intracellular trafficking in neurons. *J Biol Chem.* 2006;281(6):3552–3559.
202. Caviston JP, Ross JL, Antony SM, Tokito M, Holzbaur ELF. Huntingtin facilitates dynein/dynactin-mediated vesicle transport. *Proc Natl Acad Sci USA.* 2007;104 (24):10045–10050.
203. Godin JD, Colombo K, Molina-Calavita M, Keryer G, Zala D, Charrin BC, Dietrich P, Volvert ML, Guillemot F, Dragatsis I, Bellaiche Y, Saudou F, Nguyen L, Humbert S. Huntingtin is required for mitotic spindle orientation and mammalian neurogenesis. *Neuron.* 2010;67(3):392–406.
204. Keryer G, Pineda JR, Liot G, Kim J, Dietrich P, Benstaali C, Smith K, Cordelières FP, Spassky N, Ferrante RJ, Dragatsis I, Saudou F. Ciliogenesis is regulated by a Huntingtin-HAP1-PCM1 pathway and is altered in Huntington disease. *J Clin Invest.* 2011;121 (11):4372–4382.
205. Haremaki T, Deglincerti A, Brivanlou AH. Huntingtin is required for ciliogenesis and neurogenesis during early Xenopus development. *Dev Biol.* 2015;408(2):305–315.
206. Wong YC, Holzbaur ELF. The regulation of autophagosome dynamics by Huntingtin and HAP1 is disrupted by expression of mutant Huntingtin, leading to defective cargo degradation. *J Neurosci.* 2014;34(4):1293–1305.
207. Duyao MP, Auerbach AB, Ryan A, Persichetti F, Barnes GT, McNeil SM, Ge P, Vonsattel JP, Gusella JF, Joyner AL. Inactivation of the mouse Huntington's disease gene homolog Hdh. *Science.* 1995;269(5222):407–410.
208. Zeitlin S, Liu JP, Chapman DL, Papaioannou VE, Efstratiadis A. Increased apoptosis and early embryonic lethality in mice nullizygous for the Huntington's disease gene homologue. *Nat Genet.* 1995;11(2):155–163.
209. Nasir J, Floresco SB, O'Kusky JR, Diewert VM, Richman JM, Zeisler J, Borowski A, Marth JD, Phillips AG, Hayden MR. Targeted disruption of the Huntington's disease gene results in embryonic lethality and behavioral and morphological changes in heterozygotes. *Cell.* 1995;81(5):811–823.
210. Tong Y, Ha TJ, Liu L, Nishimoto A, Reiner A, Goldowitz D. Spatial and temporal requirements for huntingtin (HTT) in neuronal migration and survival during brain development. *J Neurosci.* 2011;31(41):14794–14799.
211. Lunkes A, Lindenberg KS, Ben-Haïem L, Weber C, Devys D, Landwehrmeyer GB, Mandel JL, Trottier Y. Proteases acting on mutant Huntingtin generate cleaved products that differentially build up cytoplasmic and nuclear inclusions. *Mol Cell.* 2002;10 (2):259–269.
212. Kim M, Roh JK, Yoon BW, Kang L, Kim YJ, Aronin N, DiFiglia M. Huntingtin is degraded to small fragments by calpain after ischemic injury. *Exp Neurol.* 2003;183 (1):109–115.
213. Gafni J, Hermel E, Young JE, Wellington CL, Hayden MR, Ellerby LM. Inhibition of calpain cleavage of Huntingtin reduces toxicity: accumulation of calpain/caspase fragments in the nucleus. *J Biol Chem.* 2004;279(19):20211–20220.

214. Kim YJ, Yi Y, Sapp E, Wang Y, Cuiffo B, Kegel KB, Qin ZH, Aronin N, DiFiglia M. Caspase 3-cleaved N-terminal fragments of wild-type and mutant Huntingtin are present in normal and Huntington's disease brains, associate with membranes, and undergo calpain-dependent proteolysis. *Proc Natl Acad Sci USA.* 2001;98 (22):12784–12789.
215. Warby SC, Doty CN, Graham RK, Carroll JB, Yang YZ, Singaraja RR, Overall CM, Hayden MR. Activated caspase-6 and caspase-6-cleaved fragments of Huntingtin specifically colocalize in the nucleus. *Hum Mol Genet.* 2008;17(15):2390–2404.
216. Kim YJ, Sapp E, Cuiffo BG, Sobin L, Yoder J, Kegel KB, Qin ZH, Detloff P, Aronin N, DiFiglia M. Lysosomal proteases are involved in generation of N-terminal Huntingtin fragments. *Neurobiol Dis.* 2006;22(2):346–356.
217. Miller JP, Holcomb J, Al-Ramahi I, de Haro M, Gafni J, Zhang N, Kim E, Sanhueza M, Torcassi C, Kwak S, Botas J, Hughes RE, Ellerby LM. Matrix metalloproteinases are modifiers of Huntingtin proteolysis and toxicity in Huntington's disease. *Neuron.* 2010;67(2):199–212.
218. Mangiarini L, Sathasivam K, Seller M, Cozens B, Harper A, Hetherington C, Lawton M, Trottier Y, Lehrach H, Davies SW, Bates GP. Exon 1 of the HD gene with an expanded CAG repeat is sufficient to cause a progressive neurological phenotype in transgenic mice. *Cell.* 1996;87(3):493–506.
219. Poirier MA, Jiang H, Ross CA. A structure-based analysis of Huntingtin mutant polyglutamine aggregation and toxicity: evidence for a compact beta-sheet structure. *Hum Mol Genet.* 2005;14(6):765–774.
220. Miller J, Arrasate M, Shaby BA, Mitra S, Masliah E, Finkbeiner S. Quantitative relationships between Huntingtin levels, polyglutamine length, inclusion body formation, and neuronal death provide novel insight into huntington's disease molecular pathogenesis. *J Neurosci.* 2010;30(31):10541–10550.
221. Graham RK, Deng Y, Slow EJ, Haigh B, Bissada N, Lu G, Pearson J, Shehadeh J, Bertram L, Murphy Z, Warby SC, Doty CN, Roy S, Wellington CL, Leavitt BR, Raymond LA, Nicholson DW, Hayden MR. Cleavage at the caspase-6 site is required for neuronal dysfunction and degeneration due to mutant Huntingtin. *Cell.* 2006;125 (6):1179–1191.
222. Zuccato C, Ciammola A, Rigamonti D, Leavitt BR, Goffredo D, Conti L, MacDonald ME, Friedlander RM, Silani V, Hayden MR, Timmusk T, Sipione S, Cattaneo E. Loss of Huntingtin-mediated BDNF gene transcription in Huntington's disease. *Science.* 2001;293(5529):493–498.
223. Gauthier LR, Charrin BC, Borrell-Pagès M, Dompierre JP, Rangone H, Cordelières FP, De Mey J, MacDonald ME, Lessmann V, Humbert S, Saudou F. Huntingtin controls neurotrophic support and survival of neurons by enhancing BDNF vesicular transport along microtubules. *Cell.* 2004;118(1):127–138.
224. Liot G, Zala D, Pla P, Mottet G, Piel M, Saudou F. Mutant Huntingtin alters retrograde transport of TrkB receptors in striatal dendrites. *J Neurosci.* 2013;33(15):6298–6309.
225. Falk JD, Vargiu P, Foye PE, Usui H, Perez J, Danielson PE, Lerner DL, Bernal J, Sutcliffe JG, Rhes:. A striatal-specific Ras homolog related to Dexras1. *J Neurosci Res.* 1999;57 (6):782–788.
226. Subramaniam S, Sixt KM, Barrow R, Snyder SH. Rhes, a striatal specific protein, mediates mutant-huntingtin cytotoxicity. *Science.* 2009;324(5932):1327–1330.
227. Swarnkar S, Chen Y, Pryor WM, Shahani N, Page DT, Subramaniam S. Ectopic expression of the striatal-enriched GTPase Rhes elicits cerebellar degeneration and an ataxia phenotype in Huntington's disease. *Neurobiol Dis.* 2015;82:66–77.
228. Parker JA, Arango M, Abderrahmane S, Lambert E, Tourette C, Catoire H, Néri C. Resveratrol rescues mutant polyglutamine cytotoxicity in nematode and mammalian neurons. *Nat Genet.* 2005;37(4):349–350.

229. Pallos J, Bodai L, Lukacsovich T, Purcell JM, Steffan JS, Thompson LM, Marsh JL. Inhibition of specific HDACs and sirtuins suppresses pathogenesis in a *Drosophila* model of Huntington's disease. *Hum Mol Genet.* 2008;17(23):3767–3775.
230. Ho DJ, Calingasan NY, Wille E, Dumont M, Beal MF. Resveratrol protects against peripheral deficits in a mouse model of Huntington's disease. *Exp Neurol.* 2010;225 (1):74–84.
231. Jeong H, Cohen DE, Cui L, Supinski A, Savas JN, Mazzulli JR, Yates JR, Bordone L, Guarente L, Krainc D. Sirt1 mediates neuroprotection from mutant huntingtin by activation of the TORC1 and CREB transcriptional pathway. *Nat Med.* 2011;18(1):159–165.
232. Jiang M, Wang J, Fu J, Du L, Jeong H, West T, Xiang L, Peng Q, Hou Z, Cai H, Seredenina T, Arbez N, Zhu S, Sommers K, Qian J, Zhang J, Mori S, Yang XW, Tamashiro KLK, Aja S, Moran TH, Luthi-Carter R, Martin B, Maudsley S, Mattson MP, Cichewicz RH, Ross CA, Holtzman DM, Krainc D, Duan W. Neuroprotective role of Sirt1 in mammalian models of Huntington's disease through activation of multiple Sirt1 targets. *Nat Med.* 2011;18(1):153–158.
233. Hathorn T, Snyder-Keller A, Messer A. Nicotinamide improves motor deficits and upregulates PGC-1α and BDNF gene expression in a mouse model of Huntington's disease. *Neurobiol Dis.* 2011;41(1):43–50.
234. Jiang M, Zheng J, Peng Q, Hou Z, Zhang J, Mori S, Ellis JL, Vlasuk GP, Fries H, Suri V, Duan W. Sirtuin 1 activator SRT2104 protects Huntington's disease mice. *Ann Clin Transl Neurol.* 2014;1(12):1047–1052.
235. Smith MR, Syed A, Lukacsovich T, Purcell J, Barbaro BA, Worthge SA, Wei SR, Pollio G, Magnoni L, Scali C, Massai L, Franceschini D, Camarri M, Gianfriddo M, Diodato E, Thomas R, Gokce O, Tabrizi SJ, Caricasole A, Landwehrmeyer B, Menalled L, Murphy C, Ramboz S, Luthi-Carter R, Westerberg G, Marsh JL. A potent and selective Sirtuin 1 inhibitor alleviates pathology in multiple animal and cell models of Huntington's disease. *Hum Mol Genet.* 2014;23(11):2995–3007.
236. Luthi-Carter R, Taylor DM, Pallos J, Lambert E, Amore A, Parker A, Moffitt H, Smith DL, Runne H, Gokce O, Kuhn A, Xiang Z, Maxwell MM, Reeves SA, Bates GP, Neri C, Thompson LM, Marsh JL, Kazantsev AG. SIRT2 inhibition achieves neuroprotection by decreasing sterol biosynthesis. *Proc Natl Acad Sci USA.* 2010;107(17):7927–7932.
237. Bobrowska A, Donmez G, Weiss A, Guarente L, Bates G. SIRT2 ablation has no effect on tubulin acetylation in brain, cholesterol biosynthesis or the progression of Huntington's disease phenotypes in vivo. *PLoS One.* 2012;7(4):e34805.
238. Chopra V, Quinti L, Kim J, Vollor L, Narayanan KL, Edgerly C, Cipicchio PM, Lauver MA, Choi SH, Silverman RB, Ferrante RJ, Hersch S, Kazantsev AG. The sirtuin 2 inhibitor AK-7 is neuroprotective in Huntington's disease mouse models. *Cell Rep.* 2012;2(6):1492–1497.
239. Quinti L, Casale M, Moniot S, Pais TF, Van Kanegan MJ, Kaltenbach LS, Pallos J, Lim RG, Naidu SD, Runne H, Meisel L, Rauf NA, Leyfer D, Maxwell MM, Saiah E, Landers JE, Luthi-Carter R, Abagyan R, Dinkova-Kostova AT, Steegborn C, Marsh JL, Lo DC, Thompson LM, Kazantsev AG. SIRT2- and NRF2-targeting thiazole-containing compound with therapeutic activity in Huntington's disease models. *Cell Chem Biol.* 2016;23(7):849–861.
240. Naia L, Rego AC, Sirtuins:. double players in Huntington's disease. *Biochim Biophys Acta.* 2015;1852(10 Pt A):2183–2194.
241. Sorolla MA, Nierga C, Rodríguez-Colman MJ, Reverter-Branchat G, Arenas A, Tamarit J, Ros J, Cabiscol E. Sir2 is induced by oxidative stress in a yeast model of Huntington disease and its activation reduces protein aggregation. *Arch Biochem Biophys.* 2011;510(1):27–34.
242. Zala D, Hinckelmann MV, Saudou F. Huntingtin's function in axonal transport is conserved in *Drosophila melanogaster.* *PLoS One.* 2013;8(3):e60162.

243. Gil JM, Rego AC. The R6 lines of transgenic mice: a model for screening new therapies for Huntington's disease. *Brain Res Rev.* 2009;59(2):410–431.
244. Schilling G, Becher MW, Sharp AH, Jinnah HA, Duan K, Kotzuk JA, Slunt HH, Ratovitski T, Cooper JK, Jenkins NA, Copeland NG, Price DL, Ross CA, Borchelt DR. Intranuclear inclusions and neuritic aggregates in transgenic mice expressing a mutant N-terminal fragment of huntingtin. *Hum Mol Genet.* 1999;8(3):397–407.
245. Cha JH, Kosinski CM, Kerner JA, Alsdorf SA, Mangiarini L, Davies SW, Penney JB, Bates GP, Young AB. Altered brain neurotransmitter receptors in transgenic mice expressing a portion of an abnormal human huntington disease gene. *Proc Natl Acad Sci USA.* 1998;95(11):6480–6485.
246. Gray M, Shirasaki DI, Cepeda C, André VM, Wilburn B, Lu XH, Tao J, Yamazaki I, Li SH, Sun YE, Li XJ, Levine MS, Yang XW. Full-length human mutant huntingtin with a stable polyglutamine repeat can elicit progressive and selective neuropathogenesis in BACHD mice. *J Neurosci.* 2008;28(24):6182–6195.
247. Hoffmann E, Wald J, Lavu S, Roberts J, Beaumont C, Haddad J, Elliott P, Westphal C, Jacobson E. Pharmacokinetics and tolerability of SRT2104, a first-in-class small molecule activator of SIRT1, after single and repeated oral administration in man. *Br J Clin Pharmacol.* 2013;75(1):186–196.
248. Hwang ES, Song SB. Nicotinamide is an inhibitor of SIRT1 in vitro, but can be a stimulator in cells. *Cell Mol Life Sci.* 2017;74(18):3347–3362.
249. Jeong H, Then F, Melia TJ, Mazzulli JR, Cui L, Savas JN, Voisine C, Paganetti P, Tanese N, Hart AC, Yamamoto A, Krainc D. Acetylation targets mutant huntingtin to autophagosomes for degradation. *Cell.* 2009;137(1):60–72.
250. Fu J, Jin J, Cichewicz RH, Hageman SA, Ellis TK, Xiang L, Peng Q, Jiang M, Arbez N, Hotaling K, Ross CA. Duan W. trans-(-)-ε-Viniferin increases mitochondrial sirtuin 3 (SIRT3), activates AMP-activated protein kinase (AMPK), and protects cells in models of Huntington Disease. *J Biol Chem.* 2012;287(29):24460–24472.
251. Trettel F, Rigamonti D, Hilditch-Maguire P, Wheeler VC, Sharp AH, Persichetti F, Cattaneo E, MacDonald ME. Dominant phenotypes produced by the HD mutation in STHdh(Q111) striatal cells. *Hum Mol Genet.* 2000;9(19):2799–2809.
252. Ramadori G, Lee CE, Bookout AL, Lee S, Williams KW, Anderson J, Elmquist JK, Coppari R. Brain SIRT1: anatomical distribution and regulation by energy availability. *J Neurosci.* 2008;28(40):9989–9996.
253. Zakhary SM, Ayubcha D, Dileo JN, Jose R, Leheste JR, Horowitz JM, Torres G. Distribution analysis of deacetylase SIRT1 in rodent and human nervous systems. *Anat Rec.* 2010;293(6):1024–1032.
254. Falone S, D'Alessandro A, Mirabilio A, Cacchio M, Di Ilio C, Di Loreto S, Amicarelli F. Late-onset running biphasically improves redox balance, energy- and methylglyoxal-related status, as well as SIRT1 expression in mouse hippocampus. *PLoS One.* 2012;7(10):e48334.
255. Revilla S, Suñol C, García-Mesa Y, Giménez-Llort L, Sanfeliu C, Cristòfol R. Physical exercise improves synaptic dysfunction and recovers the loss of survival factors in 3xTg-AD mouse brain. *Neuropharmacology.* 2014;81:55–63.
256. Quintas A, de Solís AJ, Díez-Guerra FJ, Carrascosa JM, Bogónez E. Age-associated decrease of SIRT1 expression in rat hippocampus: prevention by late onset caloric restriction. *Exp Gerontol.* 2012;47(2):198–201.
257. Wu A, Ying Z, Gomez-Pinilla F. Oxidative stress modulates Sir2alpha in rat hippocampus and cerebral cortex. *Eur J Neurosci.* 2006;23(10):2573–2580.
258. Heyward FD, Walton RG, Carle MS, Coleman MA, Garvey WT, Sweatt JD. Adult mice maintained on a high-fat diet exhibit object location memory deficits and reduced hippocampal SIRT1 gene expression. *Neurobiol Learn Mem.* 2012;98(1):25–32.

259. Cheng A, Yang Y, Zhou Y, Maharana C, Lu D, Peng W, Liu Y, Wan R, Marosi K, Misiak M, Bohr VA, Mattson MP. Mitochondrial SIRT3 mediates adaptive responses of neurons to exercise and metabolic and excitatory challenges. *Cell Metab.* 2016;23 (1):128–142.
260. Li W, Zhang B, Tang J, Cao Q, Wu Y, Wu C, Guo J, Ling EA, Liang F. Sirtuin 2, a mammalian homolog of yeast silent information regulator-2 longevity regulator, is an oligodendroglial protein that decelerates cell differentiation through deacetylating alpha-tubulin. *J Neurosci.* 2007;27(10):2606–2616.
261. Braidy N, Poljak A, Grant R, Jayasena T, Mansour H, Chan-Ling T, Smythe G, Sachdev P, Guillemin GJ. Differential expression of sirtuins in the aging rat brain. *Front Cell Neurosci.* 2015;9:167.
262. Pallàs M, Pizarro JG, Gutierrez-Cuesta J, Crespo-Biel N, Alvira D, Tajes M, Yeste-Velasco M, Folch J, Canudas AM, Sureda FX, Ferrer I, Camins A. Modulation of SIRT1 expression in different neurodegenerative models and human pathologies. *Neuroscience.* 2008;154(4):1388–1397.
263. Tulino R, Benjamin AC, Jolinon N, Smith DL, Chini EN, Carnemolla A, Bates GP. SIRT1 activity is linked to its brain region-specific phosphorylation and is impaired in huntington's disease mice. *PLoS One.* 2016;11(1). e0145425.
264. Cui L, Jeong H, Borovecki F, Parkhurst CN, Tanese N, Krainc D. Transcriptional repression of PGC-1alpha by mutant huntingtin leads to mitochondrial dysfunction and neurodegeneration. *Cell.* 2006;127(1):59–69.
265. Weydt P, Pineda VV, Torrence AE, Libby RT, Satterfield TF, Lazarowski ER, Gilbert ML, Morton GJ, Bammler TK, Strand AD, Cui L, Beyer RP, Easley CN, Smith AC, Krainc D, Luquet S, Sweet IR, Schwartz MW, La Spada AR. Thermoregulatory and metabolic defects in Huntington's disease transgenic mice implicate PGC-1alpha in Huntington's disease neurodegeneration. *Cell Metab.* 2006;4(5):349–362.
266. Brennan GP, Dey D, Chen Y, Patterson KP, Magnetta EJ, Hall AM, Dube CM, Mei YT, Baram TZ. Dual and opposing roles of MicroRNA-124 in epilepsy are mediated through inflammatory and NRSF-dependent gene networks. *Cell Rep.* 2016;14 (10):2402–2412.
267. Bano D, Zanetti F, Mende Y, Nicotera P. Neurodegenerative processes in Huntington's disease. *Cell Death Dis.* 2011;2(11):e228.
268. Labbadia J, Cunliffe H, Weiss A, Katsyuba E, Sathasivam K, Seredenina T, Woodman B, Moussaoui S, Frentzel S, Luthi-Carter R, Paganetti P, Bates GP. Altered chromatin architecture underlies progressive impairment of the heat shock response in mouse models of Huntington disease. *J Clin Invest.* 2011;121(8):3306–3319.
269. Kubota H, Kitamura A, Nagata K. Analyzing the aggregation of polyglutamine-expansion proteins and its modulation by molecular chaperones. *Methods.* 2011;53 (3):267–274.
270. Martin DDO, Ladha S, Ehrnhoefer DE, Hayden MR. Autophagy in Huntington disease and huntingtin in autophagy. *Trends Neurosci.* 2015;38(1):26–35.
271. Ashkenazi A, Bento CF, Ricketts T, Vicinanza M, Siddiqi F, Pavel M, Squitieri F, Hardenberg MC, Imarisio S, Menzies FM, Rubinsztein DC. Polyglutamine tracts regulate beclin 1-dependent autophagy. *Nature.* 2017;545(7652):108–111.
272. Evers MM, Toonen LJA, van Roon-Mom WMC. Ataxin-3 protein and RNA toxicity in spinocerebellar ataxia type 3: current insights and emerging therapeutic strategies. *Mol Neurobiol.* 2014;49(3):1513–1531.
273. Anckar J, Sistonen L. Heat shock factor 1 as a coordinator of stress and developmental pathways. *Adv Exp Med Biol.* 2007;594:78–88.
274. Perrin V, Régulier E, Abbas-Terki T, Hassig R, Brouillet E, Aebischer P, Luthi-Carter R, Déglon N. Neuroprotection by Hsp104 and Hsp27 in lentiviral-based rat models of Huntington's disease. *Mol Ther.* 2007;15(5):903–911.

275. Liu B, Larsson L, Caballero A, Hao X, Oling D, Grantham J, Nyström T. The polarisome is required for segregation and retrograde transport of protein aggregates. *Cell.* 2010;140(2):257–267.
276. Gan BQ, Tang BL. Sirt1's beneficial roles in neurodegenerative diseases—a chaperonin containing TCP-1 (CCT) connection? *Aging Cell.* 2010;9(5):924–929.
277. Lee IH, Cao L, Mostoslavsky R, Lombard DB, Liu J, Bruns NE, Tsokos M, Alt FW, Finkel T. A role for the NAD-dependent deacetylase Sirt1 in the regulation of autophagy. *Proc Natl Acad Sci USA.* 2008;105(9):3374–3379.
278. Xu L, Ash JD. The role of AMPK pathway in neuroprotection. *Adv Exp Med Biol.* 2016;854:425–430.
279. Liang Q, Benavides GA, Vassilopoulos A, Gius D, Darley-Usmar V, Zhang J. Bioenergetic and autophagic control by Sirt3 in response to nutrient deprivation in mouse embryonic fibroblasts. *Biochem J.* 2013;454(2):249–257.
280. Duan WJ, Li YF, Liu FL, Deng J, Wu YP, Yuan WL, Tsoi B, Chen JL, Wang Q, Cai SH, Kurihara H, He RR. A SIRT3/AMPK/autophagy network orchestrates the protective effects of trans-resveratrol in stressed peritoneal macrophages and RAW 264.7 macrophages. *Free Radic Biol Med.* 2016;95:230–242.
281. Inoue T, Nakayama Y, Li Y, Matsumori H, Takahashi H, Kojima H, Wanibuchi H, Katoh M, Oshimura M. SIRT2 knockdown increases basal autophagy and prevents postslippage death by abnormally prolonging the mitotic arrest that is induced by microtubule inhibitors. *FEBS J.* 2014;281(11):2623–2637.
282. Cadonic C, Sabbir MG, Albensi BC. Mechanisms of mitochondrial dysfunction in Alzheimer's disease. *Mol Neurobiol.* 2016;53(9):6078–6090.
283. Ventura-Clapier R, Garnier A, Veksler V. Transcriptional control of mitochondrial biogenesis: the central role of PGC-1alpha. *Cardiovasc Res.* 2008;79(2):208–217.
284. St-Pierre J, Drori S, Uldry M, Silvaggi JM, Rhee J, Jäger S, Handschin C, Zheng K, Lin J, Yang W, Simon DK, Bachoo R, Spiegelman BM. Suppression of reactive oxygen species and neurodegeneration by the PGC-1 transcriptional coactivators. *Cell.* 2006;127(2):397–408.
285. Kong X, Wang R, Xue Y, Liu X, Zhang H, Chen Y, Fang F, Chang Y. Sirtuin 3, a new target of PGC-1alpha, plays an important role in the suppression of ROS and mitochondrial biogenesis. *PLoS One.* 2010;5(7):e11707.
286. Hirschey MD, Shimazu T, Goetzman E, Jing E, Schwer B, Lombard DB, Grueter CA, Harris C, Biddinger S, Ilkayeva OR, Stevens RD, Li Y, Saha AK, Ruderman NB, Bain JR, Newgard CB, Farese RV, Alt FW, Kahn CR, Verdin E. SIRT3 regulates mitochondrial fatty-acid oxidation by reversible enzyme deacetylation. *Nature.* 2010;464 (7285):121–125.
287. Hirschey MD, Shimazu T, Huang JY, Schwer B, Verdin E. SIRT3 regulates mitochondrial protein acetylation and intermediary metabolism. *Cold Spring Harb Symp Quant Biol.* 2011;76:267–277.
288. Tseng AHH, Shieh SS, Wang DL. SIRT3 deacetylates FOXO3 to protect mitochondria against oxidative damage. *Free Radic Biol Med.* 2013;63:222–234.
289. Duan W, Guo Z, Jiang H, Ware M, Li XJ. Mattson MP. Dietary restriction normalizes glucose metabolism and BDNF levels, slows disease progression, and increases survival in huntingtin mutant mice. *Proc Natl Acad Sci USA.* 2003;100(5):2911–2916.
290. Nemoto S, Fergusson MM, Finkel T. SIRT1 functionally interacts with the metabolic regulator and transcriptional coactivator PGC-1{alpha}. *J Biol Chem.* 2005;280 (16):16456–16460.
291. Rodgers JT, Lerin C, Haas W, Gygi SP, Spiegelman BM, Puigserver P. Nutrient control of glucose homeostasis through a complex of PGC-1alpha and SIRT1. *Nature.* 2005;434(7029):113–118.

292. Lagouge M, Argmann C, Gerhart-Hines Z, Meziane H, Lerin C, Daussin F, Messadeq N, Milne J, Lambert P, Elliott P, Geny B, Laakso M, Puigserver P, Auwerx J. Resveratrol improves mitochondrial function and protects against metabolic disease by activating SIRT1 and PGC-1alpha. *Cell.* 2006;127(6):1109–1122.
293. Brenmoehl J, Hoeflich A. Dual control of mitochondrial biogenesis by sirtuin 1 and sirtuin 3. *Mitochondrion.* 2013;13(6):755–761.
294. Yuan Y, Cruzat VF, Newsholme P, Cheng J, Chen Y, Lu Y. Regulation of SIRT1 in aging: Roles in mitochondrial function and biogenesis. *Mech Ageing Dev.* 2016;155:10–21.
295. Tang BL. Sirt1 and the mitochondria. *Mol Cells.* 2016;39(2):87–95.
296. White JA, Anderson E, Zimmerman K, Zheng KH, Rouhani R, Gunawardena S. Huntingtin differentially regulates the axonal transport of a sub-set of Rab-containing vesicles in vivo. *Hum Mol Genet.* 2015;24(25):7182–7195.
297. Her LS, Goldstein LSB. Enhanced sensitivity of striatal neurons to axonal transport defects induced by mutant huntingtin. *J Neurosci.* 2008;28(50):13662–13672.
298. Weiss KR, Littleton JT. Characterization of axonal transport defects in *Drosophila* Huntingtin mutants. *J Neurogenet.* 2016;30(3–4):212–221.
299. Karam A, Tebbe L, Weber C, Messaddeq N, Morlé L, Kessler P, Wolfrum U, Trottier Y. A novel function of Huntingtin in the cilium and retinal ciliopathy in Huntington's disease mice. *Neurobiol Dis.* 2015;80:15–28.
300. Blum D, Herrera F, Francelle L, Mendes T, Basquin M, Obriot H, Demeyer D, Sergeant N, Gerhardt E, Brouillet E, Buée L, Outeiro TF. Mutant huntingtin alters Tau phosphorylation and subcellular distribution. *Hum Mol Genet.* 2015;24(1):76–85.
301. Gratuze M, Noël A, Julien C, Cisbani G, Milot-Rousseau P, Morin F, Dickler M, Goupil C, Bezeau F, Poitras I, Bissonnette S, Whittington RA, Hébert SS, Cicchetti F, Parker JA, Samadi P, Planel E. Tau hyperphosphorylation and deregulation of calcineurin in mouse models of Huntington's disease. *Hum Mol Genet.* 2015;24(1):86–99.
302. Fernández-Nogales M, Cabrera JR, Santos-Galindo M, Hoozemans JJM, Ferrer I, Rozemuller AJM, Hernández F, Avila J, Lucas JJ. Huntington's disease is a four-repeat tauopathy with tau nuclear rods. *Nat Med.* 2014;20(8):881–885.
303. Gratuze M, Cisbani G, Cicchetti F, Planel E. Is Huntington's disease a tauopathy? *Brain.* 2016;139(Pt 4):1014–1025.
304. Jellinger KA. Alzheimer-type lesions in Huntington's disease. *J Neural Transm.* 1998;105 (8–9):787–799.
305. Biella G, Fusco F, Nardo E, Bernocchi O, Colombo A, Lichtenthaler SF, Forloni G, Albani D. Sirtuin 2 inhibition improves cognitive performance and acts on amyloid-βprotein precursor processing in two Alzheimer's disease mouse models. *J Alzheimers Dis.* 2016;53(3):1193–1207.
306. North BJ, Marshall BL, Borra MT, Denu JM, Verdin E. The human Sir2 ortholog, SIRT2, is an NAD+-dependent tubulin deacetylase. *Mol Cell.* 2003;11(2):437–444.
307. Hong H, Kim BS, Im HI. Pathophysiological role of neuroinflammation in neurodegenerative diseases and psychiatric disorders. *Int Neurourol J.* 2016;20(Suppl 1):S2–S7.
308. Möller T. Neuroinflammation in Huntington's disease. *J Neural Transm.* 2010;117 (8):1001–1008.
309. Crotti A, Glass CK. The choreography of neuroinflammation in Huntington's disease. *Trends Immunol.* 2015;36(6):364–373.
310. Crotti A, Benner C, Kerman BE, Gosselin D, Lagier-Tourenne C, Zuccato C, Cattaneo E, Gage FH, Cleveland DW, Glass CK. Mutant Huntingtin promotes autonomous microglia activation via myeloid lineage-determining factors. *Nat Neurosci.* 2014;17 (4):513–521.
311. Hsiao HY, Chen YC, Chen HM, Tu PH, Chern Y. A critical role of astrocyte-mediated nuclear factor-κB-dependent inflammation in Huntington's disease. *Hum Mol Genet.* 2013;22(9):1826–1842.

312. Hwang JW, Yao H, Caito S, Sundar IK, Rahman I. Redox regulation of SIRT1 in inflammation and cellular senescence. *Free Radic Biol Med.* 2013;61:95–110.
313. Hayakawa T, Iwai M, Aoki S, Takimoto K, Maruyama M, Maruyama W, Motoyama N. SIRT1 suppresses the senescence-associated secretory phenotype through epigenetic gene regulation. *PLoS One.* 2015;10(1):e0116480.
314. Pais TF, Szegő ÉM. Marques O, Miller-Fleming L, Antas P, Guerreiro P, de Oliveira RM, Kasapoglu B, Outeiro TF. The NAD-dependent deacetylase sirtuin 2 is a suppressor of microglial activation and brain inflammation. *EMBO J.* 2013;32 (19):2603–2616.
315. Liddelow SA, Guttenplan KA, Clarke LE, Bennett FC, Bohlen CJ, Schirmer L, Bennett ML, Münch AE, Chung WS, Peterson TC, Wilton DK, Frouin A, Napier BA, Panicker N, Kumar M, Buckwalter MS, Rowitch DH, Dawson VL, Dawson TM, Stevens B, Barres BA. Neurotoxic reactive astrocytes are induced by activated microglia. *Nature.* 2017;541(7638):481–487.
316. Huntington Study Group, et al., Effect of deutetrabenazine on chorea among patients with Huntington disease: a randomized clinical trial. *JAMA.* 2016;316(1):40–50
317. Yang L, Calingasan NY, Wille EJ, Cormier K, Smith K, Ferrante RJ, Beal MF. Combination therapy with coenzyme Q10 and creatine produces additive neuroprotective effects in models of Parkinson's and Huntington's diseases. *J Neurochem.* 2009;109(5):1427–1439.
318. Naia L, Ribeiro MJ, Rego AC. Mitochondrial and metabolic-based protective strategies in Huntington's disease: the case of creatine and coenzyme Q. *Rev Neurosci.* 2011;23(1):13–28.
319. McGarry A, et al. A randomized, double-blind, placebo-controlled trial of coenzyme Q10 in Huntington disease. *Neurology.* 2017;88(2):152–159.
320. Shannon KM, Fraint A. Therapeutic advances in Huntington's Disease. *Mov Disord.* 2015;30(11):1539–1546.
321. Bitterman JL, Chung JH. Metabolic effects of resveratrol: addressing the controversies. *Cell Mol Life Sci.* 2015;72(8):1473–1488.
322. Turner RS, Thomas RG, Craft S, van Dyck CH, Mintzer J, Reynolds BA, Brewer JB, Rissman RA, Raman R, Aisen PS. Alzheimer's disease cooperative study. A randomized, double-blind, placebo-controlled trial of resveratrol for Alzheimer disease. *Neurology.* 2015;85(16):1383–1391.
323. Westerberg G, Chiesa JA, Andersen CA, Diamanti D, Magnoni L, Pollio G, Darpo B, Zhou M. Safety, pharmacokinetics, pharmacogenomics and QT concentration-effect modelling of the SirT1 inhibitor selisistat in healthy volunteers. *Br J Clin Pharmacol.* 2015;79(3):477–491.
324. Süssmuth SD, Haider S, Landwehrmeyer GB, Farmer R, Frost C, Tripepi G, Andersen CA, Di Bacco M, Lamanna C, Diodato E, Massai L, Diamanti D, Mori E, Magnoni L, Dreyhaupt J, Schiefele K, Craufurd D, Saft C, Rudzinska M, Ryglewicz D, Orth M, Brzozy S, Baran A, Pollio G, Andre R, Tabrizi SJ, Darpo B, Westerberg G. Paddington consortium. An exploratory double-blind, randomized clinical trial with selisistat, a SirT1 inhibitor, in patients with Huntington's disease. *Br J Clin Pharmacol.* 2015;79 (3):465–476.

CHAPTER FIVE

Mammalian Sirtuins SIRT4 and SIRT7

Shengchao Li, Weiping Zheng[1]
School of Pharmacy, Jiangsu University, Zhenjiang, People's Republic of China
[1] Corresponding author. E-mail address: wzheng@ujs.edu.cn

Contents

Abstract

Sirtuins are a family of intracellular enzymes whose enzymatic activities include catalyzing the β-nicotinamide adenine dinucleotide (β-NAD^+)-dependent N^{ε}-acyl-lysine deacylation and the β-NAD^+-dependent mono-ADP-ribosylation. Among the seven sirtuin family members (i.e., SIRT1–7) thus far identified in mammals including humans, we know SIRT1/2/3/5/6 better than SIRT4/7 as for their enzymatic activities and the cellular roles of the reactions they catalyze. This chapter will provide an updated account on the enzymology and biology of SIRT4 and SIRT7, the two less well-understood mammalian sirtuins. It is hoped that this article will also be able to set a stage for the medicinal chemistry work on SIRT4 and SIRT7, potentially developing novel therapeutic agents for human diseases.

1. SIRTUIN FAMILY OVERVIEW

The yeast and mouse transcriptional silencing silent information regulator 2 (sir2) proteins were the first sirtuin family members to be identified by virtue of their capability to catalyze the β-nicotinamide adenine dinucleotide

Progress in Molecular Biology and Translational Science, Volume 154
ISSN 1877-1173
http://dx.doi.org/10.1016/bs.pmbts.2017.11.001

(β-NAD$^+$)-dependent deacetylation of specific N^ε-acetyl-lysine side chains on histone H3 and H4 proteins.[1] Since this discovery in 2000, sir2 homologs have been found in organisms from all the three evolutionary forms of life, that is, bacteria, archaea, and eukarya.[2] In mammals, seven sirtuins (i.e., SIRT1–7) have been identified, which represents the greatest number of sirtuin family members present in an organism; in lower organisms, such as bacteria, archaea, yeast, and eukaryotic pathogens, there are fewer sirtuin family members. The ubiquitous presence of sirtuins in biosphere suggests that sirtuins are ancient proteins and participate in a variety of fundamental life processes. From what we have known so far, it is the enzymatic activity of a sirtuin that primarily underlies its regulatory role in multiple cellular processes. Specifically, sirtuins have been known to catalyze structural and thus functional modifications on an ever-increasing list of protein substrates that are present in prokaryotic cells and major intracellular compartments of eukaryotic cells and are actively involved in vital cellular processes, such as transcription, DNA damage repair, and metabolism.[3–8] As for the seven mammalian sirtuins, a multitude of native substrate proteins have been found in nucleus (including nucleolus), mitochondrion, and cytosol of a mammalian cell, consistent with the localization of the seven mammalian sirtuins in these three major intracellular compartments [nucleus (nucleoplasm): SIRT1, SIRT2, full-length SIRT3, SIRT6, SIRT7; nucleus (nucleolus): SIRT7; mitochondrion: mature SIRT3, SIRT4, SIRT5; cytosol: SIRT1, SIRT2, SIRT5, SIRT7].[3,9–13]

The sirtuin family members were initially regarded as enzymes catalyzing β-NAD$^+$-dependent N^ε-acetyl-lysine deacetylation and mono-ADP-ribosylation of proteins, however, recent investigations revealed the capability of many sirtuins to more proficiently catalyze the removal of N^ε-acyl-lysine's acyl groups bulkier than acetyl; in another word, deacetylation is actually a weak deacylation activity for many sirtuin family members.[9,14] Fig. 1 depicts the N^ε-acyl specificity of the seven human sirtuins. Moreover, mono-ADP-ribosylation of proteins has been found to be a weak activity for the majority of sirtuin family members, with the exception of a type of sirtuins from certain bacteria and fungi.[14] It should be noted that the first step of the sirtuin-catalyzed deacylation is also a mono-ADP-ribosylation reaction, just the substrate being the weak nucleophilic side chain amide oxygen of N^ε-acyl-lysine rather than stronger side chain nucleophiles (e.g., NH_2, SH) of standard amino acids on proteins.[15] Fig. 2 depicts the current version of the proposed chemical mechanisms for the sirtuin-catalyzed deacylation and mono-ADP-ribosylation. Of note, both N^ε-acyl-lysine-dependent and N^ε-acyl-lysine-independent mono-ADP-ribosylation are possible.

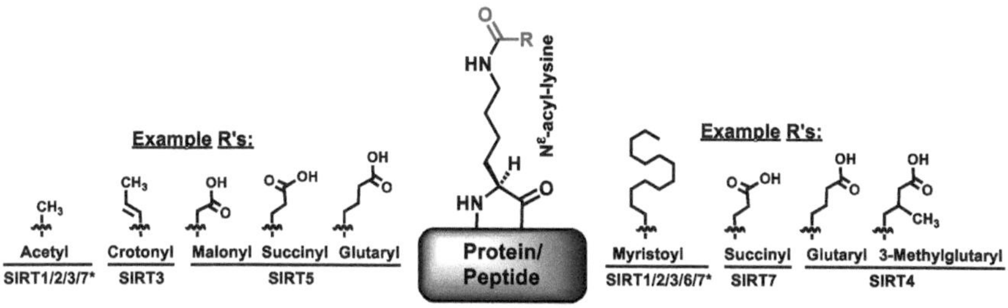

Fig. 1 The generic structure of N^{ε}-acyl-lysine in sirtuin protein/peptide substrates and the example R's whose corresponding N^{ε}-acyl groups removable by the indicated seven human sirtuins (Refs. 9,14,16–19 and see the discussion below on the N^{ε}-acyl specificity of SIRT4 and SIRT7). The *asterisk* indicates that the SIRT7-catalyzed deacetylation and demyristoylation become appreciable only in the presence of dsDNA, rRNA, or tRNA (Refs. 17 and 18 and see the discussion further).

Another major advancement in the sirtuin field over past few years is the elucidation of the 3-dimensional structures for sirtuins from organisms in all the three evolutionary domains of life.[27,28] The free sirtuins and their complexes with substrates/inhibitors have both yielded to structural elucidation. The availability of such structural information has been instrumental in enhancing our mechanistic understanding of the sirtuin-catalyzed deacylation reaction,[9,20] and this information has been gradually exploited for a structure-based design of inhibitors for the sirtuin-catalyzed deacylation reaction.[20] Fig. 3 illustrates the two-domain topology of the conserved sirtuin catalytic core, alongside with the domain organization of the seven full-length human sirtuins.

As for the structure/function studies on the mammalian sirtuins SIRT1–7, SIRT1 has been the most extensively studied, followed by SIRT2, SIRT3, SIRT5, and SIRT6. SIRT4 and SIRT7 have been the least well-understood, however, the studies on these two sirtuins have been increasingly reported during the past few years. In the following sections we will elaborate on the two major profiles of these two sirtuins, that is, their respective enzymatic activities and cellular functions enabled by their enzymatic actions.

2. SIRT4 ENZYMATIC ACTIVITIES

The first enzymatic activity discovered for SIRT4 was its β-NAD^+-dependent mono-ADP-ribosylation and inhibition of the mitochondrial enzyme glutamate dehydrogenase in pancreatic β cells, which resulted in

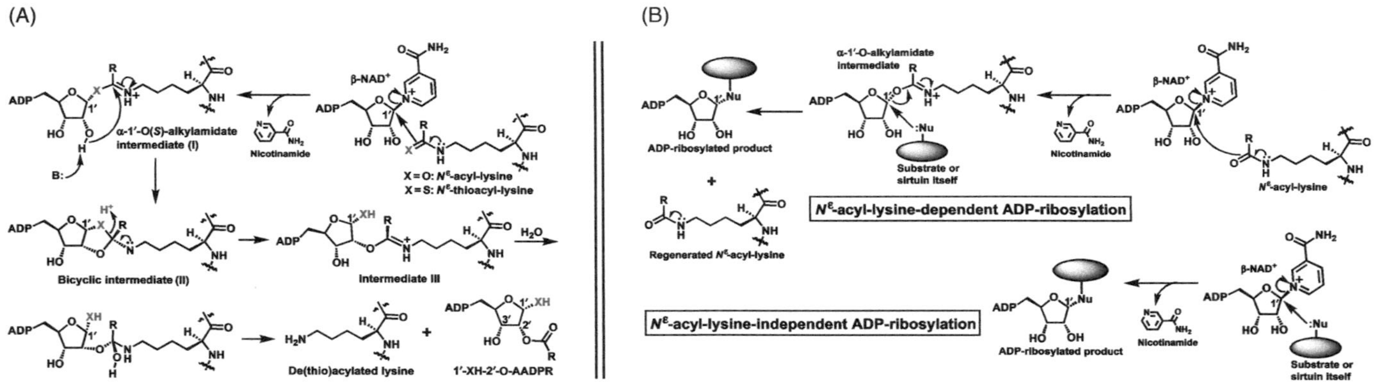

Fig. 2 The current version of the proposed chemical mechanisms for the sirtuin-catalyzed deacylation [(A), X = O] and mono-ADP-ribosylation (B). Note: (1) The sirtuin-catalyzed dethioacylation [(A), X = S] is also shown, given the great contribution of the use of N^{ε}-thioacyl-lysines to the elucidation of the chemical mechanism of sirtuin-catalyzed deacylation, primarily via solving the X-ray cocrystal structures of sirtuins with the longer-lived thio version of intermediates I, II, and III [(A), X = S] trapped within their active sites.[9,20,21] However, with regard to the depicted route for the resolution of the bicyclic intermediate II in the presence of water to afford the two end products, which has been proposed based on the trapping of the thio version of intermediate III [(A), X = S, R = tridecyl] within SIRT2 active site,[21] in our opinion, this proposal needs one more thought. Specifically, since thiolate ($-S^-$) is a better leaving group than alkoxide ($-O^-$), C—S bond can be cleaved in the absence of a general acid while abstracting a proton from bulk reaction solution, however, C—O bond cleavage would be more possible with the help of a general acid. In this regard, it would be critical to see if there is such a general acid appropriately positioned in the solved cocrystal SIRT2 structure. If there is no such a general acid, then the observed formation of intermediate III may be unique to the use of the thioamide-harboring N^{ε}-thiomyristoyl-lysine; in line with this reasoning, the SIRT2 (or sirtuin in general) deacylation may not go through intermediate III (X = O) and the downstream catalytic intermediate with the N^{ε}-acyl-lysine substrate. (2) Both N^{ε}-acyl-lysine-dependent and N^{ε}-acyl-lysine-independent mono-ADP-ribosylation could be possible (B), based on the studies on the mono-ADP-ribosylation activity of several sirtuins.[22–26] *AADPR, acyl-ADP-ribose; Nu*, nucleophile.

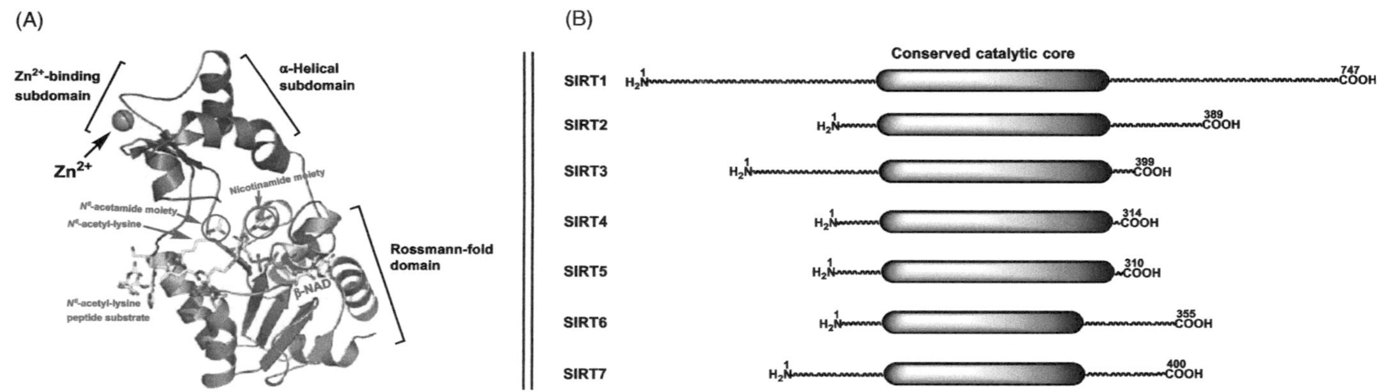

Fig. 3 (A) The two-domain topology of the conserved sirtuin catalytic core, as illustrated with the experimentally determined 3-dimensional structure of the bacterial sirtuin Sir2Tm complexed with the N^{ε}-acetyl-lysine substrate peptide and β-NAD^+.[29] The two domains in this structure refer to the larger Rossmann-fold domain and a smaller domain composed of the α-helical subdomain and the Zn^{2+}-binding subdomain. The N^{ε}-acetyl-lysine substrate and β-NAD^+ are rendered with the stick model and are shown to both bind at the interface of the two domains. (B) The domain organization of the seven full-length human sirtuins. The conserved sirtuin catalytic core is indicated with the *gray* bar. *Part A, This structural view was adapted with permission from Hoff KG, Avalos JL, Sens K, Wolberger C. Insights into the sirtuin mechanism from ternary complexes containing NAD+ and acetylated peptide. Structure. 2006;14:1231–1240*[29]*; part B, modified from Fig. 4 in Chen B, Zang W, Wang J, Huang Y, He Y, Yan L, Liu J, Zheng W. The chemical biology of sirtuins. Chem Soc Rev. 2015;44:5246–5264.*[9]

downregulation of the amino acid-stimulated secretion of insulin.[30] Subsequently, SIRT4 was shown to be also likely able to mono-ADP-ribosylate insulin degrading enzyme and ATP/ADP translocases ANT2 and ANT3 (transmembrane proteins across the inner mitochondrial membrane) in pancreatic β cells, downregulating the glucose-stimulated insulin secretion.[31]

SIRT4 was also shown to catalyze the N^{ε}-acetyl-lysine deacetylation of the mitochondrial malonyl-CoA decarboxylase which is an enzyme catalyzing the conversion of malonyl-CoA to acetyl-CoA.[32] This deacetylation was also shown to result in an inhibition of the enzymatic activity of malonyl-CoA decarboxylase and a downregulation of the fatty acid oxidation in muscle cells (see further).

SIRT4 was further found to be able to catalyze the β-NAD^+-dependent delipoylation of N^{ε}-lipoyl-lysine on the E2 component dihydrolipoyl-lysine acetyltransferase of mitochondrial pyruvate dehydrogenase complex and thus to inhibit the activity of this complex.[33] This delipoylase (or lipoamidase) activity of SIRT4 was found in the study to be more proficient than its deacetylase activity. Moreover, it is intriguing that SIRT4-catalyzed delipoylation of pyruvate dehydrogenase complex can be promoted in the event of a metabolic flux switch to glutamine providing the carbon source of citric acid cycle, highlighting the important role of SIRT4 in regulating metabolism via inhibiting the activity of pyruvate dehydrogenase complex. In this same study, SIRT4 was also found to possess a debiotinylase activity even though the native substrate(s) of this activity remain(s) to be identified.

It should be noted that the aforementioned β-NAD^+-dependent mono-ADP-ribosylation, deacetylation, and delipoylation activities of SIRT4 are all weak.[14] However, a very recent study found that SIRT4 was able to more proficiently catalyze the β-NAD^+-dependent removal of the glutaryl and 3-methylglutaryl groups respectively from N^{ε}-glutaryl-lysine and N^{ε}-(3-methylglutaryl)-lysine.[16] Since the metabolic intermediates responsible for donating the acyl groups of these two N^{ε}-acyl-lysine residues were derived from the leucine oxidation pathway, it is not surprising to note that the SIRT4 deacylase activities identified in the study were indeed found to be involved in the regulation of leucine metabolism, specifically, the activation of leucine oxidation. It was also found in the study that SIRT4 played an important role in regulating insulin secretion and glucose homeostasis, since deletion of SIRT4 in a mouse model promoted insulin secretion ultimately leading to glucose intolerance and insulin resistance. This finding could be accounted for by the SIRT4 deletion-induced diminished leucine oxidation and thus the promoted leucine-stimulated insulin secretion.[34] In a subsequent study by the same research group,[35] it was further found that the enhanced leucine-stimulated insulin secretion in SIRT4 knockout mice was genetic background-dependent. Moreover, SIRT4 knockout was found in the study to only marginally affect lipid metabolism.

3. SIRT4 CELLULAR FUNCTIONS

3.1 SIRT4 in Metabolism

The mitochondrial SIRT4 is intimately involved in metabolic regulation, which includes the above-described down regulation of insulin secretion in pancreatic β cells, fatty acid oxidation in liver and muscle cells.[30,36,37] This is consistent with the expression pattern of SIRT4 both at the mRNA and protein levels. By Northern blot analysis, SIRT4 was found to be expressed in human muscle, kidney, testis, and liver cells,[38] whereas by an immunohistochemistry analysis using an anti-SIRT4 antibody, SIRT4 was found to be significantly expressed in vascular smooth muscle cells and skeletal muscle cells, and the insulin-producing pancreatic β cells; this last localization was confirmed by the immunohistochemistry analysis using an antiinsulin antibody.[31] In the study, SIRT4 was also found to be weakly expressed in pancreatic acini.

The inhibitory effect of SIRT4 against fatty acid oxidation in muscle cells, as described above, was recently shown to be mediated by the SIRT4-catalyzed deacetylation and activity inhibition of the mitochondrial malonyl-CoA decarboxylase, an enzyme that catalyzes the conversion of malonyl-CoA to acetyl-CoA.[32] Since malonyl-CoA and acetyl-CoA are the two starting metabolites for fatty acid biosynthesis, the SIRT4-catalyzed deacetylation and the ensuing activity decrease of malonyl-CoA decarboxylase would lead to an accumulation of malonyl-CoA and the ensuing increased fatty acid biosynthesis. Moreover, since malonyl-CoA inhibits carnitine palmitoyltransferase I (located on the outer mitochondrial membrane) that helps to transport the cytosolic fatty acyl-CoA into mitochondrial matrix for the subsequent β-oxidation, the accumulation of malonyl-CoA would downregulate fatty acid oxidation. Therefore, the SIRT4-catalyzed deacetylation of malonyl-CoA decarboxylase simultaneously promotes fatty acid anabolism and inhibits fatty acid catabolism.

Recently, the SIRT4-mediated downregulation of fatty acid oxidation in liver cells was found to be mediated by the suppressive effect of SIRT4 against the transcriptional activity of the nuclear receptor peroxisome proliferator-activated receptor α (PPARα).[39] Mechanistically, SIRT4 interferes with the binding interaction of SIRT1 with PPARα, therefore attenuating the activation effect of SIRT1 on the transcriptional activity of PPARα and inhibiting fatty acid oxidation, as supported by the SIRT4 knockout and overexpression experiments on cellular and animal models. It should be noted that, under normal condition in liver cells, via interacting with PPARα, SIRT1 gets recruited to the PPARα response element (PPRE) where it catalyzes the N^{ϵ}-acetyl-lysine deacetylation of the transcriptional coactivator peroxisome proliferator-activated receptor gamma coactivator 1-alpha (PGC-1α), thereby promoting fatty acid oxidation.[40] Moreover, the usage of β-NAD^+ by the SIRT4 enzymatic activities would make β-NAD^+ less available for other sirtuins including SIRT1, therefore leading to reduced SIRT1 overall activity and further attenuated activation of SIRT1 on the transcriptional activity of PPARα and fatty acid oxidation. The inhibitory effect of SIRT4 on liver fatty acid oxidation can also explain the observed SIRT4 level decrease in liver cells during fasting[39] since fatty acid oxidation would be promoted to be used as a source of cellular energy fuel during fasting. In this study, SIRT4 was also established to be involved in mitochondrial retrograde signaling, a mode of communication of mitochondria with the nucleus for cells to maintain metabolic homeostasis via adjusting nuclear transcriptional activities in response to cellular energy status fluctuation.[41]

In another recent study, SIRT4 was shown to regulate not only fatty acid oxidation but also cellular ATP homeostasis and mitochondrial biogenesis.[42] SIRT4 deletion decreases cellular ATP level while SIRT4 overexpression increases cellular ATP level. This impact of SIRT4 on cellular ATP homeostasis was found to be mediated by the ATP/ADP translocase 2 (ANT2, transmembrane protein across the inner mitochondrial membrane). Mechanistically, in addition to the aforementioned mono-ADP-ribosylation, SIRT4 may also catalyze a N^{ϵ}-acyl-lysine deacylation on ANT2 so as to dampen the mitochondrial uncoupling leading to an enhanced ATP production. It should be noted that ANT2 lysine N^{ϵ}-acylation was found previously to promote mitochondrial uncoupling, therefore leading to an enhanced mitochondrial respiration yet a decreased

ATP production.[43–45] The ANT2-mediated effect of SIRT4 on cellular ATP level was shown in the study also to be a part of a mitochondrial retrograde signaling pathway. The low ATP level due to SIRT4 deficiency activates AMP-activated protein kinase (AMPK, a cytosol/nucleus-localized regulator of cellular energy homeostasis) and the activated AMPK in turn would phosphorylate and inhibit the cytosolic acetyl-CoA carboxylase (ACC) that catalyzes the conversion of acetyl-CoA to malonyl-CoA, leading to a reduction of the malonyl-CoA level in cytosol. Since malonyl-CoA is an inhibitor of carnitine palmitoyltransferase I (located on the outer mitochondrial membrane) that helps to transport the cytosolic fatty acyl-CoA into mitochondrial matrix for the subsequent β-oxidation, as aforementioned, the decreased abundance of malonyl-CoA would then promote fatty acid oxidation. Moreover, the low ATP level and the activated AMPK in response to SIRT4 deficiency was also shown in the study to increase the expression of the transcriptional coactivator PGC-1α and its target genes involved in fatty acid oxidation and mitochondrial biogenesis, therefore also leading to a promoted fatty acid oxidation in addition to a promoted mitochondrial biogenesis.

3.2 SIRT4 in Cancer

SIRT4 has been found to possess tumor suppressive effect by virtue of its metabolic regulatory role, consistent with the importance of mitochondrial metabolism in tumorigenesis. The identification of SIRT4 as a mitochondrion-localized tumor suppressor protein has also expanded the notion that tumor suppressor proteins can be present in not only nucleus and cytosol but also in mitochondrion.[46]

SIRT4 was recently found to be an important player in the metabolic response to DNA damage by mediating the DNA damage-induced inhibition of glutamine catabolism.[47,48] Since glutamine can be catabolized by mitochondrial glutaminase to form glutamate followed by the generation of the citric acid cycle intermediate α-ketoglutarate from glutamate by the reactions catalyzed by mitochondrial glutamate dehydrogenase and aspartate aminotransferase, the SIRT4-mediated inhibition of glutamine catabolism, specifically the anaplerotic replenishment into citric acid cycle, would create a cellular state for DNA damage repair. This SIRT4-mediated inhibition of glutamine catabolism would also lead to a tumor suppressive effect since it has an antiproliferative effect against those cells harboring DNA damages. The tumor suppressive effect of SIRT4 was also supported by further experimental observations in the study.[48] Specifically, while SIRT4 expression was found to be upregulated by DNA damage, its expression was found to be downregulated in many types of human cancer cells; SIRT4 deletion was found to promote the glutamine-dependent cell proliferation and lead to genomic instability in response to genotoxic stress; SIRT4 knockout mice can develop tumors.

Consistent with the above-described tumor suppressive role of SIRT4 via inhibiting mitochondrial glutamine catabolism, another recent study found that SIRT4 was also able to suppress the growth of the transcription factor c-Myc-induced B cell lymphoma, which could be attributed to the SIRT4-mediated repression of the enhanced glutamine catabolism induced by the aberrant activation of c-Myc and the essential role of glutamine catabolism for the c-Myc-induced cancers.[49] In the study, SIRT4 overexpression in c-Myc-driven human Burkitt lymphoma cells was found to result in repression of glutamine catabolism and the glutamine-dependent cell proliferation, as well as a synergism with glycolysis inhibitors in promoting cell death. On the other hand, in a mouse model of c-Myc-driven Burkitt lymphoma, SIRT4 knockout was shown to exhibit an accelerated lymphomagenesis and mortality, as well as an enhanced glutamine uptake and the enzymatic activity of glutamate dehydrogenase. The authors of the study further found that the tumor suppressive role of SIRT4 in the c-Myc-induced B cell lymphoma was independent of c-Myc itself.

Another recent study[50] employing SIRT4-overexpressing human colorectal cancer cell lines found that SIRT4 overexpression increased the expression of E-cadherin, leading to a suppression of cell proliferation, migration, and invasion. However, this phenotypic consequence was a result of SIRT4-mediated inhibition of glutamine catabolism. It was further found in this study that SIRT4 expression became lower with the progression of colorectal cancer invasion and metastasis.

The tumor suppressive effect of SIRT4 in colorectal cancer was also reported in still another recent study.[51] Specifically, from a survey of 89 colorectal cancer samples, SIRT4 expression was found to be significantly lower than that in the corresponding normal tissue samples. Moreover, colorectal cancer cell proliferation and the glutamine catabolism inside were also found to be inhibited by SIRT4 overexpression, with the latter inhibition being found to be synergistic with glycolysis inhibitors to induce cell death. SIRT4 was also found to increase the sensitivity of colorectal cancer to chemotherapeutic agent 5-fluorouracil via an inhibition of the cell cycle, consistent with the antiproliferative effect of SIRT4 overexpression.

4. SIRT7 ENZYMATIC ACTIVITIES AND CELLULAR FUNCTIONS

While the cellular functions of SIRT7 have been excellently reviewed recently,[52] here we would like to discuss the cellular functions of SIRT7 side-by-side with its enzymatic activities since the SIRT7 cellular functions are primarily enabled by its enzymatic activities. Before this discussion, we would like to use the following example to illustrate the existence of the SIRT7 cellular functions independent of its enzymatic activity.

In a recent study,[53] SIRT7 was found to inhibit the transcriptional activity of transcription factors hypoxia-inducible factor (HIF)-1 and HIF-2 via decreasing the protein levels of their respective subunits HIF-1α and HIF-2α. This effect of SIRT7 was shown to be independent of its enzymatic activity since treatment with high concentrations of the broad-spectrum sirtuin inhibitor nicotinamide, or mutation of the catalytic residues or deletion of its catalytic core, had no impact on its inhibitory action. Even though the mechanism of this inhibitory action of SIRT7 needs to be established, it is possible that SIRT7 indirectly regulates the protein levels of HIF-1α and HIF-2α via protein–protein interactions.

In terms of the enzymatic activities of SIRT7, it was first found to be a β-NAD$^+$-dependent deacetylase enzyme able to catalyze the side chain acetyl removal from the N^{ε}-acetyl-lysine residue at position 18 of histone H3 protein on chromatin.[54] This deacetylation was found to occur after SIRT7 gets recruited to the promoters of specific target genes (e.g., tumor suppressor genes) via its interaction with the E26 transformed specific (ETS) domain-containing transcription factor ELK4, thus repressing the expression of these genes and promoting oncogenic transformation. In the study, it was also demonstrated that this deacetylation is highly selective toward the 18th position versus other tested positions on histone H3 protein, and versus the tested lysine N^{ε}-acetylatable positions on H4, H2A, and H2B proteins.

In a later study,[11] this SIRT7-catalyzed deacetylation was shown to be regulatable via controlling the intracellular spatial distribution of the enzyme (i.e., SIRT7) by Dicer protein. Specifically, it was found in the study that human Dicer protein was able to interact with SIRT7 and consequently keeps a fraction of SIRT7 within cytoplasm, which was supported by the Dicer knockdown and overexpression experiments. As a result, the SIRT7-catalyzed deacetylation in nucleus of histone H3 protein at position 18 was diminished. The findings in the study that DNA damaging agent treatments induced the expression of Dicer protein and the consequent decrease in SIRT7-catalyzed deacetylation of histone H3 protein at position 18 suggested that, in the presence of genotoxic stress, the oncogenic role of this SIRT7-catalyzed deacetylation would be normally suppressed.

SIRT7 was found in another later study to be also able to catalyze the β-NAD^+-dependent N^{ε}-acetyl-lysine deacetylation on a nonhistone substrate.[10] Specifically, when binding to RNA polymerase I (Pol I) and its transcription factor upstream binding factor (UBF), SIRT7 is able to catalyze the side chain acetyl removal from the N^{ε}-acetyl-lysine residue at position 373 of the Pol I-associated factor PAF53 (a subunit of Pol I), thus promoting the binding of Pol I with the negatively charged ribosomal DNA (rDNA) and the rDNA transcription. It was found in the study that the rDNA transcription facilitated by this deacetylation could also be subjected to a regulation via the control over the intracellular spatial distribution of SIRT7. For example, under cellular stress conditions, SIRT7 was found to be translocated from nucleolus to nucleoplasm, therefore SIRT7 became no longer available for the deacetylation of PAF53 and the hyperacetylated PAF53 rendered Pol I less competent in catalyzing the pre-rRNA synthesis due to its diminished binding capacity to the negatively charged rDNA. Moreover, the diminished pre-rRNA synthesis would reduce the retention of SIRT7 in nucleolus, which requires its binding with the nascent pre-rRNA. Further, the binding of SIRT7 with the nascent pre-rRNA was also shown to mediate the interaction of SIRT7 with its substrate PAF53 (and thus Pol I).

In addition to regulating the rDNA transcription with the synthesis of pre-rRNA, as described above, the same research group later on reported that SIRT7 also played an important role in regulating the processing and maturation of pre-rRNA via catalyzing the N^{ε}-acetyl-lysine deacetylation on another nonhistone protein, that is, U3-55k protein which is a key protein component in the U3 small nucleolar RNP (snoRNP) complex.[55] Specifically, the SIRT7-catalyzed deacetylation of U3-55k at positions 12 and 25 promotes the binding interaction of U3-55k with the negatively charged U3 snoRNA (an RNA component in the U3 snoRNP complex), which is required for the processing of pre-rRNA. Therefore, as observed in the study, the cellular stress-induced translocation of SIRT7 from nucleolus to nucleoplasm and the consequent decreased deacetylation of U3-55k by SIRT7 attenuated the processing of pre-rRNA.

Besides the above-described roles of SIRT7 in regulating the Pol I-mediated rDNA transcription and the processing and maturation of pre-rRNA, SIRT7 was also shown by the same research group to play an important role in regulating the transcription mediated by the RNA polymerase II (Pol II).[56] Specifically, SIRT7 was found in the study to catalyze the N^{ε}-acetyl-lysine deacetylation and thus the kinase activity activation of the cyclin-dependent kinase 9 (CDK9) at position 48 within the kinase domain of CDK9. Since CDK9 is a part of the positive transcription elongation factor P-TEFb, the CDK9 deacetylation also resulted in the release and activation of P-TEFb from the 7SK small nuclear ribonucleoprotein (snRNP) complex. The activated P-TEFb subsequently phosphorylates (due to its activated CDK9) the Ser2 residue at the carboxyl terminal domain of Pol II, hence promoting the Pol II-mediated transcription.

A recent study suggested that SIRT7 might also play an important role in regulating the transcription mediated by the RNA polymerase III (Pol III).[57] Specifically, SIRT7 was found to interact with mammalian target of rapamycin (mTOR), and SIRT7 and mTOR in the resulting binary complex can both interact with TFIIIC2, one transcription factor complex for Pol III, thus potentially impact the transcription mediated by Pol III with the synthesis of transfer RNA (tRNA) and the translation process. As a support, SIRT7 knockdown was shown in the study to decrease the tRNA synthesis. However, the mechanistic detail for the activation of Pol III transcription by SIRT7, particularly concerning the role of SIRT7 enzymatic activity in this activating action, is yet to be elucidated.

Still another nonhistone substrate found for SIRT7 is GA binding protein β1 (GABPβ1), a transcription factor whose deacetylation of N^{ε}-acetyl-lysine residues by SIRT7 at positions 69, 340, and 369 mediates the regulation of the transcription of multiple nuclear-encoded mitochondrial genes.[58] It was found in the study that SIRT7 deficiency caused mitochondrial functional defects such that there was an increased blood lactate concentration, the occurrence of hepatic microvesicular steatosis, etc.

In one recent study,[59] damage-specific DNA binding protein 1 (DDB1) in the cullin 4B (CUL4B)/DDB1/DDB1-CUL4-associated factor 1 (DCAF1) E3 ubiquitin ligase complex was found to be a nonhistone substrate of SIRT7. The SIRT7-catalyzed N^{ε}-acetyl-lysine deacetylation of DDB1 at position 1121 was supposed to be able to attenuate the binding interaction of DDB1 with DCAF1 (another protein component in the CUL4B/DDB1/DCAF1 E3 ubiquitin ligase complex), as supported by the experiment on the N^{ε}-acetyl-lysine deacetylation-mimicking mutant DDB1[K1121R]. Moreover, the attenuated binding interaction of the N^{ε}-acetyl-lysine deacetylated DDB1 with DCAF1 would decrease the ligase activity of the CUL4B/DDB1/DCAF1 E3 ubiquitin ligase complex that was found previously by the authors to be responsible for degrading the nuclear receptor testicular receptor 4 (TR4),[60] therefore there would then be an increase in the TR4 protein level and the expression of its target genes Cd36, Cidea, Cidec, and Pparg1. Consequently, there would be an increased lipid synthesis and storage in liver. The findings from this study have furthered the mechanistic understanding of the resistance to fatty liver development in response to a high-fat diet by SIRT7 knockout mice the authors previously studied.[60] It should be noted that the observed resistance to fatty liver development in response to a high-fat diet by SIRT7 knockout mice in this previous study by Yoshizawa *et al.* is in stark contrast to the observed promotion of fatty liver development by SIRT7 knockout in another two studies by Ryu *et al.* and Shin *et al.*, respectively.[58,61] For a detailed analysis of these contradictory observations, the readers are referred to an excellent recent Opinion Article.[62] As pointed out in this Opinion Article, despite these discrepancies, SIRT7 undoubtedly plays an important role in regulating hepatic lipid metabolism, which ought to be also consistent with the previous observation that liver is a major tissue where SIRT7 is expressed, as judged by the mRNA (by Northern blot) and protein (by Western blot) levels.[63] Here within the general frame of our discussion, we would also like to point out that the SIRT7-catalyzed N^{ε}-acetyl-lysine deacetylation plays an important mechanistic role in the regulation of hepatic lipid metabolism by SIRT7. Specifically, besides the above-described mechanistic roles of the SIRT7-catalyzed deacetylation of DDB1 and GABPβ1, the SIRT7-catalyzed deacetylation on histone H3 protein at position 18 was suggested in the study by Shin *et al.* to also play an important mechanistic role in regulating hepatic lipid metabolism via controlling the expression of ribosomal protein genes.[61] Specifically, SIRT7 gets recruited to the promoters of ribosomal proteins via its interaction with transcription factor Myc and there catalyzes the deacetylation on histone H3 protein at position 18, leading to suppression of gene expression, endoplasmic reticulum (ER) stress, and fatty liver development. On the other hand, SIRT7 knockout was observed in mice to promote fatty liver development, which can be reverted by SIRT7 expression.

As suggested from the above studies, the SIRT7-catalyzed deacetylation obviously plays an important role in regulating histone and nonhistone substrate-mediated cellular processes, however, SIRT7 was also found previously not to possess a proficient in vitro deacetylase activity when evaluated in cell-free assay with purified substrate and SIRT7,[64] suggesting that SIRT7 is a proficient deacetylase only when it and/or its substrate are complexed with other molecules, for example, the aforementioned rRNA and the nucleosomal DNA. Within this context, two recent studies have systematically explored the in vitro conditions that would make SIRT7 a proficient deacetylase.[17,18] Specifically, in the first study, with added double-stranded DNA (dsDNA) in an in vitro activity assay based on high performance liquid chromatography, SIRT7 was found to be able to proficiently catalyze the acetyl removal from the N^{ε}-acetyl-lysine residue at position 18 of a histone H3 peptide, in contrast to a lack of detectable deacetylation product in the absence of dsDNA. Moreover, in a Western blot-based in vitro assay, SIRT7 was found to be unable to catalyze the acetyl removal from the N^{ε}-acetyl-lysine residue at position 18 of the purified histone H3 protein regardless of whether dsDNA was added to the assay system, however, SIRT7 was able to proficiently catalyze the acetyl removal from the N^{ε}-acetyl-lysine residue at positions 9 and 18 of histone H3 protein which was present in nucleosome. This finding suggests that SIRT7 can bind to and be activated by the nucleosomal DNA and achieve an efficient catalytic removal of the acetyl group from the acetylated core histone H3 protein also present in the same complex. In the second study, the same researchers found that RNA species (including rRNA and tRNA) could enhance the in vitro SIRT7 deacetylase activity to a greater extent than dsDNA. Moreover, in the presence of rRNA or tRNA, SIRT7 was found to more proficiently catalyze the myristoyl removal from N^{ε}-myristoyl-lysine than the acetyl removal from N^{ε}-acetyl-lysine in vitro. It was also found in the study via metabolic labeling that the RNA-dependent SIRT7 defatty-acylase activity is physiologically relevant. Taken together, via defining the in vitro conditions for proficient SIRT7 deacylase activities, the findings from these two studies would facilitate the discovery of chemical modulators (inhibitors and activators) for SIRT7.

In another recent study,[19] SIRT7 was found to be recruited to DNA double-strand break (DSB) sites in a poly (ADP-ribose) polymerase 1 (PARP-1)-dependent manner and catalyze the β-NAD^{+}-dependent on-site succinyl removal from the N^{ε}-succinyl-lysine residue at position 122 of histone H3 protein, thus facilitating the chromatin condensation and DNA DSB damage repair. In the study, SIRT7 was found to be much more proficient in catalyzing this desuccinylation than in catalyzing the deacetylation at position 122 of histone H3 protein. However, the SIRT7-catalyzed desuccinylation was found in an in vitro assay not to be specific for position 122 of histone H3 protein, it could also occur with high proficiency at positions 56 and 79 of histone H3 protein. Therefore, the cellular effect of SIRT7-catalyzed desuccinylation could be mediated by the desuccinylation at not only position 122 but also positions 56 and 79 of histone H3 protein. Moreover, it is interesting to note that the SIRT7-catalyzed desuccinylation at position 122 of histone H3 protein in mononucleosomes (harboring dsDNA) and calf thymus histones was found in the study to exhibit a comparable proficiency, a scenario unlike the above-described SIRT7-catalyzed more proficient deacetylation on histone H3 protein within the chromatin context than on the purified histone H3 protein.

Apart from the above-described role of the SIRT7-catalyzed desuccinylation in promoting the genome integrity and DNA DSB repair, the SIRT7-catalyzed deacetylation at position 18 of histone H3 protein was recently reported to play a similar role.[65] Specifically, the recruitment of SIRT7 in a PARP-1–dependent manner to the DNA DSB sites would promote the N^{ε}-acetyl-lysine deacetylation at position 18 of histone H3 protein. And this deacetylation would then help to recruit the damage response factor 53BP1 to the DNA DSB sites to promote the DSB repair by nonhomologous end joining. In the study, the authors also observed that SIRT7 deletion in mice resulted in embryonic lethality and the development of a progeroid-like phenotype. And a compromised DNA damage repair and an increased replication stress were observed with the cells deficient in SIRT7.

From the above discussion, the currently known endogenous SIRT7 deacylase substrates include the Lys18-acetylated histone H3 protein, the Lys122-succinylated histone H3 protein, the Lys373-acetylated PAF53 protein, the Lys12/25-acetylated U3-55k protein, the Lys69/340/369-acetylated GABPβ1 protein, the Lys1121-acetylated DDB1 protein, and the Lys48-acetylated CDK9 protein. It should be noted that the major tumor suppressor protein p53 was found to be a SIRT7 deacetylase substrate in an in vitro assay only;[66] and nucleophosmin (NPM1) is a N^{ε}-acetyl-lysine deacetylation substrate shared by SIRT7 with SIRT6, another nuclear sirtuin.[67] The identification of the above native substrates has been instrumental to an enhanced mechanistic/functional understanding of various cellular roles of SIRT7 enabled by its N^{ε}-acyl-lysine deacylase activities. In an effort to expand the SIRT7 deacetylase substrate repertoire, a quantitative proteomics approach was taken recently.[68,69] In the study, stable isotope labeling with amino acids in cell culture (SILAC) and quantitative mass spectrometry were performed with SIRT7$^{+/+}$ and SIRT7$^{-/-}$ mouse embryonic fibroblasts, and 261 acetylation sites in 176 proteins were identified as candidate SIRT7 deacetylation sites and substrates. Many of these candidate SIRT7 deacetylase substrates were predicted to be involved in metabolic processes *per* a bioinformatics-based analysis of the mass spectrometry data. Moreover, the SIRT7-catalyzed deacetylation of a few selected candidate substrates was also confirmed with in vitro deacetylation assay. The findings from this study suggest that SIRT7 may accept many proteins as its native deacetylase substrates.

5. CONCLUSION AND PERSPECTIVE

In this chapter, we have provided an updated account on the enzymatic activities that have been identified for SIRT4 and SIRT7, the two less well-understood mammalian sirtuins, and their respective cellular functions that are enabled by their enzymatic activities. Specifically, while SIRT4 was found to catalyze a β-NAD^+-dependent mono-ADP-ribosylation reaction, SIRT4 and SIRT7 are both able to catalyze the β-NAD^+-dependent side chain deacylation of N^ε-acyl-lysines whose acyl groups include acetyl and bulkier ones. Moreover, SIRT4 and SIRT7 are both involved in metabolic regulation. However, while SIRT4 is generally regarded as a tumor suppressor, SIRT7 is a tumor promoter. Given this knowledge of these two sirtuins and with a hope of potentially developing novel therapeutic agents for human diseases via targeting the enzymatic reactions, particularly the deacylation reactions, catalyzed by these two sirtuins, one future challenge will be how to develop effective (potent, selective, metabolically stable, and cell permeable) chemical modulators (inhibitors or activators) for them. These chemical modulators could also be employed as chemical biology research tools for a further elucidation of the cellular functions of SIRT4 and SIRT7. These chemical modulators could further be employed as pharmacological research tools for a further clarification of the therapeutic potential following inhibiting or activating the SIRT4 (or SIRT7)-catalyzed reactions, which in the case with SIRT7 would complement the extensively reported genetic inhibition/inactivation (by RNAi, gene deletion, or microRNA), gene overexpression, and SIRT7 mRNA and protein level profiling in various cancers suggesting that oncogenic SIRT7 constitutes a novel anticancer therapeutic target.[70–78]

Since there has been only one structural study in the current literature for these two sirtuins, that is, the solution of a 2.3 Å X-ray crystal structure for the N-terminal domain of human SIRT7 fused to maltose-binding protein (MBP);[79] in another word, there has been no any structural information available for the full-length or the catalytic core of SIRT4 and SIRT7, the structure-based lead inhibitor (or activator) discovery (via virtual screening) would be unrealistic to undertake for now, neither would be the ligand-based lead discovery due to the limited number of the (un)natural ligands currently discovered/developed for these two sirtuins. However, the lead discovery based on chemical library screening could be successful, given the above-described proficient deacylase activities for SIRT4 and SIRT7. The catalytic mechanism-based design of the lead inhibitors (or activators) for SIRT4 and SIRT7 could be also successful, given the catalytic mechanism conservation across the sirtuin family members.[9,20] This latter approach would also enjoy its simplicity and effectiveness, as already documented extensively in literature (for lead inhibitor design in particular).[9,20] Developing the catalytic mechanism-based SIRT4 and SIRT7 inhibitors also constitutes an ongoing effort in our laboratory.

Another future challenge will be to expand our repertoire of native substrates for SIRT4 and SIRT7. To this end, the above-described recent study on SIRT7 by Zhang *et al.*[68] would serve as a valuable starting point. On the other hand, the peptide microarray approach to narrow down candidate sirtuin substrates, as reported by Rauh *et al.*,[80] could also be helpful. The findings from this effort would also expand our functional horizon of SIRT4 and SIRT7.

ACKNOWLEDGMENTS

The work on sirtuins in W.Z.'s laboratory has been supported by the following: National Natural Science Foundation of China (grant no. 21272094), the Jiangsu Provincial Specially Appointed Professorship, the Jiangsu Provincial "Innovation and Venture Talents" award plan, Jiangsu University, the US National Institutes of Health (CA152972), the James L. and Martha J. Foght Endowment, and University of Akron.

REFERENCES

1. Imai S, Armstrong CM, Kaeberlein M, Guarente L. Transcriptional silencing and longevity protein Sir2 is an NAD-dependent histone deacetylase. *Nature.* 2000;403:795–800.
2. Greiss S, Gartner A. Sirtuin/Sir2 phylogeny, evolutionary considerations and structural conservation. *Mol Cells.* 2009;28:407–415.
3. Martínez-Redondo P, Vaquero A. The diversity of histone versus nonhistone sirtuin substrates. *Genes Cancer.* 2013;4:148–163.
4. Carabetta VJ, Cristea IM. Regulation, function, and detection of protein acetylation in bacteria. *J Bacteriol.* 2017;199:e00107–e00117.
5. Elkhwanky MS, Hakkola J. Extranuclear sirtuins and metabolic stress. *Antioxid Redox Signal.* 2017.
6. Dai Y, Faller DV. Transcription regulation by class III histone deacetylases (HDACs)-sirtuins. *Transl Oncogenomics.* 2008;3:53–65.

7. Choi JE, Mostoslavsky R. Sirtuins, metabolism, and DNA repair. *Curr Opin Genet Dev.* 2014;26:24–32.
8. Sebastián C, Mostoslavsky R. The role of mammalian sirtuins in cancer metabolism. *Semin Cell Dev Biol.* 2015;43:33–42.
9. Chen B, Zang W, Wang J, Huang Y, He Y, Yan L, Liu J, Zheng W. The chemical biology of sirtuins. *Chem Soc Rev.* 2015;44:5246–5264.
10. Chen S, Seiler J, Santiago-Reichelt M, Felbel K, Grummt I, Voit R. Repression of RNA polymerase I upon stress is caused by inhibition of RNA-dependent deacetylation of PAF53 by SIRT7. *Mol Cell.* 2013;52:303–313.
11. Zhang PY, Li G, Deng ZJ, Liu LY, Chen L, Tang JZ, Wang YQ, Cao ST, Fang YX, Wen F, Xu Y, Chen X, Shi KQ, Li WF, Xie C, Tang KF. Dicer interacts with SIRT7 and regulates H3K18 deacetylation in response to DNA damaging agents. *Nucleic Acids Res.* 2016;44:3629–3642.
12. Kiran S, Chatterjee N, Singh S, Kaul SC, Wadhwa R, Ramakrishna G. Intracellular distribution of human SIRT7 and mapping of the nuclear/nucleolar localization signal. *FEBS J.* 2013;280:3451–3466.
13. Matsushita N, Yonashiro R, Ogata Y, Sugiura A, Nagashima S, Fukuda T, Inatome R, Yanagi S. Distinct regulation of mitochondrial localization and stability of two human Sirt5 isoforms. *Genes Cells.* 2011;16:190–202.
14. Bheda P, Jing H, Wolberger C, Lin H. The substrate specificity of sirtuins. *Annu Rev Biochem.* 2016;85:405–429.
15. Palazzo L, Mikoč A, Ahel I. ADP-ribosylation: new facets of an ancient modification. *FEBS J.* 2017;284:2932–2946.
16. Anderson KA, Huynh FK, Fisher-Wellman K, Stuart JD, Peterson BS, Douros JD, Wagner GR, Thompson JW, Madsen AS, Green MF, Sivley RM, Ilkayeva OR, Stevens RD, Backos DS, Capra JA, Olsen CA, Campbell JE, Muoio DM, Grimsrud PA, Hirschey MD. SIRT4 is a lysine deacylase that controls leucine metabolism and insulin secretion. *Cell Metab.* 2017;25:838–855.
17. Tong Z, Wang Y, Zhang X, Kim DD, Sadhukhan S, Hao Q, Lin H. SIRT7 is activated by DNA and deacetylates histone H3 in the chromatin context. *ACS Chem Biol.* 2016;11:742–747.
18. Tong Z, Wang M, Wang Y, Kim DD, Grenier JK, Cao J, Sadhukhan S, Hao Q, Lin H. SIRT7 is an RNA-activated protein lysine deacylase. *ACS Chem Biol.* 2017;12:300–310.
19. Li L, Shi L, Yang S, Yan R, Zhang D, Yang J, He L, Li W, Yi X, Sun L, Liang J, Cheng Z, Shi L, Shang Y, Yu W. SIRT7 is a histone desuccinylase that functionally links to chromatin compaction and genome stability. *Nat Commun.* 2016;7:12235.
20. Jiang Y, Liu J, Chen D, Yan L, Zheng W. Sirtuin inhibition: strategies, inhibitors, and therapeutic potential. *Trends Pharmacol Sci.* 2017;38:459–472.
21. Wang Y, Fung YM, Zhang W, He B, Chung MW, Jin J, Hu J, Lin H, Hao Q. Deacylation mechanism by SIRT2 revealed in the 1'-SH-2'-O-myristoyl intermediate structure. *Cell Chem Biol.* 2017;24:339–345.
22. Hawse WF, Wolberger C. Structure-based mechanism of ADP-ribosylation by sirtuins. *J Biol Chem.* 2009;284:33654–33661.
23. Kowieski TM, Lee S, Denu JM. Acetylation-dependent ADP-ribosylation by Trypanosoma brucei Sir2. *J Biol Chem.* 2008;283:5317–5326.
24. Fahie K, Hu P, Swatkoski S, Cotter RJ, Zhang Y, Wolberger C. Side chain specificity of ADP-ribosylation by a sirtuin. *FEBS J.* 2009;276:7159–7176.
25. French JB, Cen Y, Sauve AA. Plasmodium falciparum Sir2 is an NAD^+-dependent deacetylase and an acetyllysine-dependent and acetyllysine-independent NAD^+ glycohydrolase. *Biochemistry.* 2008;47:10227–10239.
26. Du J, Jiang H, Lin H. Investigating the ADP-ribosyltransferase activity of sirtuins with NAD analogues and ^{32}P-NAD. *Biochemistry.* 2009;48:2878–2890.

27. Sacconnay L, Carrupt PA, Nurisso A. Human sirtuins: structures and flexibility. *J Struct Biol.* 2016;196:534–542.
28. Yuan H, Marmorstein R. Structural basis for sirtuin activity and inhibition. *J Biol Chem.* 2012;287:42428–42435.
29. Hoff KG, Avalos JL, Sens K, Wolberger C. Insights into the sirtuin mechanism from ternary complexes containing NAD^+ and acetylated peptide. *Structure.* 2006;14: 1231–1240.
30. Haigis MC, Mostoslavsky R, Haigis KM, Fahie K, Christodoulou DC, Murphy AJ, Valenzuela DM, Yancopoulos GD, Karow M, Blander G, Wolberger C, Prolla TA, Weindruch R, Alt FW, Guarente L. SIRT4 inhibits glutamate dehydrogenase and opposes the effects of calorie restriction in pancreatic beta cells. *Cell.* 2006;126:941–954.
31. Ahuja N, Schwer B, Carobbio S, Waltregny D, North BJ, Castronovo V, Maechler P, Verdin E. Regulation of insulin secretion by SIRT4, a mitochondrial ADP-ribosyltransferase. *J Biol Chem.* 2007;282:33583–33592.
32. Laurent G, German NJ, Saha AK, de Boer VC, Davies M, Koves TR, Dephoure N, Fischer F, Boanca G, Vaitheesvaran B, Lovitch SB, Sharpe AH, Kurland IJ, Steegborn C, Gygi SP, Muoio DM, Ruderman NB, Haigis MC. SIRT4 coordinates the balance between lipid synthesis and catabolism by repressing malonyl CoA decarboxylase. *Mol Cell.* 2013;50:686–698.
33. Mathias RA, Greco TM, Oberstein A, Budayeva HG, Chakrabarti R, Rowland EA, Kang Y, Shenk T, Cristea IM. Sirtuin 4 is a lipoamidase regulating pyruvate dehydrogenase complex activity. *Cell.* 2014;159:1615–1625.
34. Hutton JC, Sener A, Malaisse WJ. Interaction of branched chain amino acids and keto acids upon pancreatic islet metabolism and insulin secretion. *J Biol Chem.* 1980;255:7340–7346.
35. Huynh FK, Hu X, Lin Z, Johnson JD, Hirschey MD. Loss of sirtuin 4 leads to elevated glucose- and leucine-stimulated insulin levels and accelerated age-induced insulin resistance in multiple murine genetic backgrounds. *J Inherit Metab Dis.* 2017.
36. Nasrin N, Wu X, Fortier E, Feng Y, Bare' OC, Chen S, Ren X, Wu Z, Streeper RS, Bordone L. SIRT4 regulates fatty acid oxidation and mitochondrial gene expression in liver and muscle cells. *J Biol Chem.* 2010;285:31995–32002.
37. Argmann C, Auwerx J. Insulin secretion: SIRT4 gets in on the act. *Cell.* 2006;126: 837–839.
38. Shi T, Wang F, Stieren E, Tong Q. SIRT3, a mitochondrial sirtuin deacetylase, regulates mitochondrial function and thermogenesis in brown adipocytes. *J Biol Chem.* 2005;280:13560–13567.
39. Laurent G, de Boer VC, Finley LW, Sweeney M, Lu H, Schug TT, Cen Y, Jeong SM, Li X, Sauve AA, Haigis MC. SIRT4 represses peroxisome proliferator-activated receptor α activity to suppress hepatic fat oxidation. *Mol Cell Biol.* 2013;33:4552–4561.
40. Purushotham A, Schug TT, Xu Q, Surapureddi S, Guo X, Li X. Hepatocyte-specific deletion of SIRT1 alters fatty acid metabolism and results in hepatic steatosis and inflammation. *Cell Metab.* 2009;9:327–338.
41. Finley LW, Haigis MC. The coordination of nuclear and mitochondrial communication during aging and calorie restriction. *Ageing Res Rev.* 2009;8:173–188.
42. Ho L, Titus AS, Banerjee KK, George S, Lin W, Deota S, Saha AK, Nakamura K, Gut P, Verdin E, Kolthur-Seetharam U. SIRT4 regulates ATP homeostasis and mediates a retrograde signaling via AMPK. *Aging (Albany NY).* 2013;5:835–849.
43. Klingenberg M. The ADP and ATP transport in mitochondria and its carrier. *Biochim Biophys Acta.* 2008;1778:1978–2021.
44. Brustovetsky N, Klingenberg M. The reconstituted ADP/ATP carrier can mediate H^+ transport by free fatty acids, which is further stimulated by mersalyl. *J Biol Chem.* 1994;269:27329–27336.

45. Forman NG, Wilson DF. Dependence of mitochondrial oxidative phosphorylation on activity of the adenine nucleotide translocase. *J Biol Chem.* 1983;258:8649–8655.
46. Zhu Y, Yan Y, Principe DR, Zou X, Vassilopoulos A, Gius D. SIRT3 and SIRT4 are mitochondrial tumor suppressor proteins that connect mitochondrial metabolism and carcinogenesis. *Cancer Metab.* 2014;2:15.
47. Fernandez-Marcos PJ, Serrano M. Sirt4: the glutamine gatekeeper. *Cancer Cell.* 2013;23: 427–428.
48. Jeong SM, Xiao C, Finley LW, Lahusen T, Souza AL, Pierce K, Li YH, Wang X, Laurent G, German NJ, Xu X, Li C, Wang RH, Lee J, Csibi A, Cerione R, Blenis J, Clish CB, Kimmelman A, Deng CX, Haigis MC. SIRT4 has tumor-suppressive activity and regulates the cellular metabolic response to DNA damage by inhibiting mitochondrial glutamine metabolism. *Cancer Cell.* 2013;23:450–463.
49. Jeong SM, Lee A, Lee J, Haigis MC. SIRT4 protein suppresses tumor formation in genetic models of Myc-induced B cell lymphoma. *J Biol Chem.* 2014;289:4135–4144.
50. Miyo M, Yamamoto H, Konno M, Colvin H, Nishida N, Koseki J, Kawamoto K, Ogawa H, Hamabe A, Uemura M, Nishimura J, Hata T, Takemasa I, Mizushima T, Doki Y, Mori M, Ishii H. Tumour-suppressive function of SIRT4 in human colorectal cancer. *Br J Cancer.* 2015;113:492–499.
51. Huang G, Cheng J, Yu F, Liu X, Yuan C, Liu C, Chen X, Peng Z. Clinical and therapeutic significance of sirtuin-4 expression in colorectal cancer. *Oncol Rep.* 2016;35:2801–2810.
52. Blank MF, Grummt I. The seven faces of SIRT7. *Transcription.* 2017;8:67–74.
53. Hubbi ME, Hu H, Kshitiz. Gilkes DM, Semenza GL. Sirtuin-7 inhibits the activity of hypoxia-inducible factors. *J Biol Chem.* 2013;288:20768–20775.
54. Barber MF, Michishita-Kioi E, Xi Y, Tasselli L, Kioi M, Moqtaderi Z, Tennen RI, Paredes S, Young NL, Chen K, Struhl K, Garcia BA, Gozani O, Li W, Chua KF. SIRT7 links H3K18 deacetylation to maintenance of oncogenic transformation. *Nature.* 2012;487:114–118.
55. Chen S, Blank MF, Iyer A, Huang B, Wang L, Grummt I, Voit R. SIRT7-dependent deacetylation of the U3-55k protein controls pre-rRNA processing. *Nat Commun.* 2016;7:10734.
56. Blank MF, Chen S, Poetz F, Schnölzer M, Voit R, Grummt I. SIRT7-dependent deacetylation of CDK9 activates RNA polymerase II transcription. *Nucleic Acids Res.* 2017;45:2675–2686.
57. Tsai YC, Greco TM, Cristea IM. Sirtuin 7 plays a role in ribosome biogenesis and protein synthesis. *Mol Cell Proteomics.* 2014;13:73–83.
58. Ryu D, Jo YS, Lo Sasso G, Stein S, Zhang H, Perino A, Lee JU, Zeviani M, Romand R, Hottiger MO, Schoonjans K, Auwerx J. A SIRT7-dependent acetylation switch of GABPβ1 controls mitochondrial function. *Cell Metab.* 2014;20:856–869.
59. Karim MF, Yoshizawa T, Sobuz SU, Sato Y, Yamagata K. Sirtuin 7-dependent deacetylation of DDB1 regulates the expression of nuclear receptor TR4. *Biochem Biophys Res Commun.* 2017;490:423–428.
60. Yoshizawa T, Karim MF, Sato Y, Senokuchi T, Miyata K, Fukuda T, Go C, Tasaki M, Uchimura K, Kadomatsu T, Tian Z, Smolka C, Sawa T, Takeya M, Tomizawa K, Ando Y, Araki E, Akaike T, Braun T, Oike Y, Bober E, Yamagata K. SIRT7 controls hepatic lipid metabolism by regulating the ubiquitin-proteasome pathway. *Cell Metab.* 2014;19:712–721.
61. Shin J, He M, Liu Y, Paredes S, Villanova L, Brown K, Qiu X, Nabavi N, Mohrin M, Wojnoonski K, Li P, Cheng HL, Murphy AJ, Valenzuela DM, Luo H, Kapahi P, Krauss R, Mostoslavsky R, Yancopoulos GD, Alt FW, Chua KF, Chen D. SIRT7 represses Myc activity to suppress ER stress and prevent fatty liver disease. *Cell Rep.* 2013;5:654–665.
62. Tang BL. SIRT7 and hepatic lipid metabolism. *Front Cell Dev Biol.* 2015;3:1.

63. Ford E, Voit R, Liszt G, Magin C, Grummt I, Guarente L. Mammalian Sir2 homolog SIRT7 is an activator of RNA polymerase I transcription. *Genes Dev.* 2006;20:1075–1080.
64. Feldman JL, Baeza J, Denu JM. Activation of the protein deacetylase SIRT6 by long-chain fatty acids and widespread deacylation by mammalian sirtuins. *J Biol Chem.* 2013;288:31350–31356.
65. Vazquez BN, Thackray JK, Simonet NG, Kane-Goldsmith N, Martinez-Redondo P, Nguyen T, Bunting S, Vaquero A, Tischfield JA, Serrano L. SIRT7 promotes genome integrity and modulates non-homologous end joining DNA. repair. *EMBO J.* 2016;35:1488–1503.
66. Vakhrusheva O, Smolka C, Gajawada P, Kostin S, Boettger T, Kubin T, Braun T, Bober E. Sirt7 increases stress resistance of cardiomyocytes and prevents apoptosis and inflammatory cardiomyopathy in mice. *Circ Res.* 2008;102:703–710.
67. Lee N, Kim DK, Kim ES, Park SJ, Kwon JH, Shin J, Park SM, Moon YH, Wang HJ, Gho YS, Choi KY. Comparative interactomes of SIRT6 and SIRT7: implication of functional links to aging. *Proteomics.* 2014;14:1610–1622.
68. Zhang C, Zhai Z, Tang M, Cheng Z, Li T, Wang H, Zhu WG. Quantitative proteome-based systematic identification of SIRT7 substrates. *Proteomics.* 2017;17(13–14).
69. Simonet NG, Vaquero A. Raising the list of SirT7 targets to a new level. *Proteomics.* 2017;17(13–14).
70. Ashraf N, Zino S, Macintyre A, Kingsmore D, Payne AP, George WD, Shiels PG. Altered sirtuin expression is associated with node-positive breast cancer. *Br J Cancer.* 2006;95:1056–1061.
71. Han Y, Liu Y, Zhang H, Wang T, Diao R, Jiang Z, Gui Y, Cai Z. Hsa-miR-125b suppresses bladder cancer development by down-regulating oncogene SIRT7 and oncogenic long non-coding RNA MALAT1. *FEBS Lett.* 2013;587:3875–3882.
72. Yu H, Ye W, Wu J, Meng X, Liu RY, Ying X, Zhou Y, Wang H, Pan C, Huang W. Overexpression of sirt7 exhibits oncogenic property and serves as a prognostic factor in colorectal cancer. *Clin Cancer Res.* 2014;20:3434–3445.
73. Zhang S, Chen P, Huang Z, Hu X, Chen M, Hu S, Hu Y, Cai T. Sirt7 promotes gastric cancer growth and inhibits apoptosis by epigenetically inhibiting miR-34a. *Sci Rep.* 2015;5:9787.
74. Malik S, Villanova L, Tanaka S, Aonuma M, Roy N, Berber E, Pollack JR, Michishita-Kioi E, Chua KF. SIRT7 inactivation reverses metastatic phenotypes in epithelial and mesenchymal tumors. *Sci Rep.* 2015;5:9841.
75. Shi H, Ji Y, Zhang D, Liu Y, Fang P. MicroRNA-3666-induced suppression of SIRT7 inhibits the growth of non-small cell lung cancer cells. *Oncol Rep.* 2016;36:3051–3057.
76. Kim JK, Noh JH, Jung KH, Eun JW, Bae HJ, Kim MG, Chang YG, Shen Q, Park WS, Lee JY, Borlak J, Nam SW. Sirtuin7 oncogenic potential in human hepatocellular carcinoma and its regulation by the tumor suppressors MiR-125a-5p and MiR-125b. *Hepatology.* 2013;57:1055–1067.
77. Aljada A, Saleh AM, Alkathiri M, Shamsa HB, Al-Bawab A, Nasr A. Altered sirtuin 7 expression is associated with early stage breast cancer. *Breast Cancer (Auckl).* 2015;9:3–8.
78. Wang HL, Lu RQ, Xie SH, Zheng H, Wen XM, Gao X, Guo L. SIRT7 exhibits oncogenic potential in human ovarian cancer cells. *Asian Pac J Cancer Prev.* 2015;16:3573–3577.
79. Priyanka A, Solanki V, Parkesh R, Thakur KG. Crystal structure of the N-terminal domain of human SIRT7 reveals a three-helical domain architecture. *Proteins.* 2016;84:1558–1563.
80. Rauh D, Fischer F, Gertz M, Lakshminarasimhan M, Bergbrede T, Aladini F, Kambach C, Becker CF, Zerweck J, Schutkowski M, Steegborn C. An acetylome peptide microarray reveals specificities and deacetylation substrates for all human sirtuin isoforms. *Nat Commun.* 2013;4:2327.

INDEX

O

P

T

U

X

Y

FSC
www.fsc.org
MIX
Papier | Fördert gute Waldnutzung
FSC® C083411

Kolibri 360 GmbH
Ferdinand-Jühlke-Straße 7
99095 Erfurt, Deutschland
produktsicherheit@kolibri360.de